# Thinking about Oneself

# Thinking about Oneself

From Nonconceptual Content to the Concept of a Self

Kristina Musholt

The MIT Press
Cambridge, Massachusetts
London, England

MIT Press books may be purchased at special quantity discounts for business or sales promotional use. For information, please e-mail special_sales@mitpress.mit.edu.

This book was set in Stone by the MIT Press. Printed and bound in the United States of America.

Library of Congress Cataloging-in-Publication Data

Musholt, Kristina.
Thinking about oneself : from nonconceptual content to the concept of self / Kristina Musholt.
   pages   cm
Includes bibliographical eferences and index.
ISBN 978-0-262-02920-9 (hardcover : alk. paper)
1. Self (Philosophy) 2. Self-consciousness (Awareness) 3. Intersubjectivity. I. Title.
BD450.M885   2015
126—dc23
                                                    2015001164

10  9  8  7  6  5  4  3  2  1

# Contents

# Acknowledgments

This book has been in the making for a long time. Its writing would not have been possible without the support of numerous individuals and institutions. I apologize in advance to anyone whose name I might accidentally omit from these acknowledgments.

The book is based on my doctoral dissertation, "Self-Consciousness: From Nonconceptual Content to the Concept of a Self," which I defended in 2011 at the Humboldt University of Berlin. I am grateful to my dissertation advisers, Michael Pauen and Ulman Lindenberger, for their encouragement and support, as well as for numerous helpful discussions and feedback on my work. Tobias Rosefeldt also deserves thanks for his comments on my dissertation. The Studienstiftung des deutschen Volkes and the Berlin School of Mind and Brain provided financial as well as intellectual support during my doctoral studies.

From 2007 to 2009, I was fortunate enough to spend eighteen months at the Department of Linguistics and Philosophy at the Massachusetts Institute of Technology in Cambridge. I am indebted to the entire department, and in particular to my host, Alex Byrne, for their generosity in hosting me and in sharing their knowledge and enthusiasm for philosophy, and for valuable feedback on my work.

During my time in Cambridge, I also had the pleasure of working as a teaching assistant with Jerry Samet at Brandeis University and with Owen Flanagan at a Harvard University summer school. I am grateful to Jerry and Owen for everything that I learned from them and for their continuous support since. I am especially grateful to Owen for helping to make this project a reality.

While reworking the material from my dissertation into this book, I benefited greatly from the collegiality and support provided by the members of the Department of Philosophy, Logic, and Scientific Method at the London School of Economics and the Forum for European Philosophy, where I worked from 2010 to 2013, and by the members of the Philosophy Department at the University of Magdeburg, where I took up a position in 2013. Material that went into this book was presented at a number of departmental talks, conferences, and workshops, including at the University of Münster, the University of Leipzig, the University of Magdeburg, the University of Parma, the Max Planck Institute for Human Development in Berlin, the Northern Institute of Philosophy, the University of Luxembourg, the University of Edinburgh, and Cardiff University; ESPP meetings in Budapest, London, and Granada; a workshop on immunity to error through misidentification in Konstanz; the London School of Economics; the Forum for European Philosophy; the Institute of Philosophy in London; the University of Koblenz-Landau; Cambridge University; the Philosophy of Mind and Language conference in Stockholm; the ECAP meeting in Milan; ASSC meetings in Taipei and Berlin; a Mind Group workshop in Berlin; the CUNY Graduate Center in New York; the Humboldt University of Berlin; and the University of Bochum. I am grateful to the audiences at these meetings for their helpful and constructive comments and suggestions.

In addition to those already mentioned, several colleagues deserve thanks for reading parts of or even the entire manuscript at various stages, providing valuable comments and encouragement, or influencing my thinking about the topic of self-consciousness through conversations. Many thanks to Arnon Cahen, Glenn Carruthers, Katja Crone, James Dow, Malte Engel, Anika Fiebich, Jörg Fingerhut, Daniel Friedrich, Marie Guillot, Ginger Hoffman, Tomis Kapitan, Andrea Lailach-Hennrich, Roblin Meeks, Jan Prause-Stamm, Marisa Przyrembel, François Recanati, Kranti Saran, Anna Strasser, and Gottfried Vosgerau. I also received extremely generous, challenging and constructive comments on the manuscript from no fewer than six anonymous readers for the MIT Press, which did much to improve the manuscript. Felicia Höer deserves thanks for help with the references. Needless to say, all remaining errors are my own.

I would also like to thank my editor, Phil Laughlin, for believing in the project and for his professionalism, patience, and support throughout the process; Judy Feldmann and Christopher Eyer, for their professional and efficient management of the editing process; and William G. Henry, for his excellent copyediting.

Finally, I am grateful for the support received from my family and friends. I owe my greatest thanks to Patrick Wilken, who was there for me from the very beginning to the project's end, read and commented on several versions of the manuscript over the years and supported the project in numerous other ways. Without him, this book would not exist.

## Permissions

Chapter 3 draws heavily—with kind permission from Springer Press—from my article "Self-Consciousness and Nonconceptual Content," *Philosophical Studies* 163 (2013): 649–672, copyright Springer Science and Business Media BV 2011.

Chapter 6 is partly based on my article "Self-Consciousness and Intersubjectivity," *Grazer Philosophische Studien* 84 (2012): 63–89.

# Introduction

Self-consciousness, that is, the ability to think "I"-thoughts, is a topic that traditionally takes center stage in many philosophical theories. More recently, it has also entered into the focus of empirical investigation in psychology and neuroscience. This is unsurprising, for the ability to think about ourselves lies at the core of what makes us human. It lies at the root of the special concern that each of us has for ourselves and enables us, among other things, to reflect on ourselves and our life trajectories and to ask questions of moral significance. After all, I can only start to think about an experience, a thought, an action, or a character trait of mine if I recognize it *as* being mine. It is by being able to think about myself that I can develop a sense of the kind of person I am, make plans for the future and reflect on past actions, or ask questions such as what life I want to lead, and what choices I should make to pursue the kind of life I regard as worth living. This ability is also often thought to separate humans from nonhuman animals—though, as we will see later in the book, the question of whether and to what extent nonhuman animals share with us the ability to think about themselves is difficult to answer.

Self-consciousness is also at the center of this book, whose aim it is to provide a solution to what I call the problem of self-consciousness. As already indicated, self-consciousness can be understood as the ability to think "I"-thoughts. As will be seen in detail in chapter 1, the problem of self-consciousness consists of the challenge of providing an account of this ability that is neither circular nor reliant on mysterious elements that cannot be further analyzed. To meet this challenge, I investigate the relevance of recent developments in the philosophy of cognition with regard to the discussion of so-called nonconceptual forms of representation, as well as the relevance of recent empirical findings, particularly those in developmental psychology. I argue that theories of self-consciousness that are based on theories of non-conceptual content, while being on the right track, are incomplete either because they fail to distinguish between implicitly self-related information and explicit self-representation (in the case of what I call self-representationalist theories, such as the theory defended by Bermúdez, 1998) or because they leave open how to make intelligible the transition from implicitly self-related information to explicit self-representation (in the case of what I call non-self-representationalist, or "no-self," theories, such as those defended by Peacocke, 1999; Recanati, 2007; and O'Brien, 2007).[1] I propose a model that explains the gradual transition from self-related information that is implicit in the nonconceptual content of perception and bodily experience to the explicit representation of the self in conceptual thought, based on a process of "representational redescription" (Karmiloff-Smith, 1996). A crucial part of this model will consist in an analysis of the relationship between self-consciousness and intersubjectivity.

Before I present an overview of how I will address the problem of self-consciousness, a few conceptual clarifications are in order. First, let me state what this book will *not* be dealing with. This book will not deal with the problem of consciousness. So I will not address issues related to "qualia" or the "explanatory gap" problem (although I might on occasion discuss issues that are of some relevance to theories of consciousness, in particular in chapter 2). While these are deep and important issues that have sparked a huge amount of debate, they are not in the focus of this book. Rather, what I aim to examine is how a representational state can be about the subject entertaining it, such that the subject forms an "I"-thought, marking the subject as an individual entity in the world and subject of the thought in question. What I cannot and do not want to explain is how a state becomes conscious in the first place, in the sense of "what-it-is-like" for the subject. I will take the consciousness of (some) mental states for granted as a necessary condition for self-consciousness, so to speak, to focus on the specific representational characteristics that make some of these conscious states *self*-conscious states—and thus distinguish self-conscious creatures from those that are merely conscious.

Second, let me briefly explain what I mean by self-consciousness (I will go into more detail in chapter 1). As mentioned before, self-consciousness as I understand the term is the ability to think "I"-thoughts. In analogy to a distinction that is familiar in debates about consciousness, I suggest that we should distinguish between state self-consciousness and creature self-consciousness. We can say of a creature that it is conscious in the sense of being sentient, and we can say of a particular mental state that it is conscious, such as the conscious perception of a color, or the conscious experience of the taste of coffee (Rosenthal, 1986). A creature that is conscious must, in principle, be in a position to entertain conscious mental states (sensations, emotions, thoughts, etc.), though it may not, at any given time, entertain any (for instance, because it is sleeping). That is to say that a creature that is conscious must, in principle, be able to be the subject of conscious mental states. Similarly, we can say of a creature that it is self-conscious if that creature is in principle able to entertain self-conscious states. As we will see in more detail in chapter 1, self-conscious states are those that would typically be expressed by means of the first-person pronoun, as in "I am hungry," "I see a tree in front of me," or "I hope Germany will win the World Cup."[2] Hence, insofar as a creature is in principle able to entertain representational states whose content has the form of "I"-thoughts, this creature is endowed with self-consciousness. My aim in this book is to provide an account of the kind of representational capacities a being must have to be able to entertain such thoughts.

Choosing a capacities-based approach has the advantage of avoiding from the outset the kind of Cartesianism that is characteristic of positions that tend to conceive of the self as a kind of substance that defies naturalistic explanation. Moreover, a capacities-based approach lends itself more easily to the project of operationalizing the relevant capacities, which is a prerequisite to their empirical investigation. This is important because it allows for a fruitful cooperation between philosophy and the empirical sciences, allowing each discipline to benefit from the insights gained by the other. Accordingly, my approach can be seen as a broadly naturalistic approach in the sense of such a mutually beneficial exchange. However, this should not be mistaken for a reductionist approach. I do not think that philosophical concepts such as self-consciousness can be reduced to subpersonal computational processes or neuronal firing patterns. Nonetheless philosophy does not—or in any case should not—operate in a vacuum and, where relevant, should take into account empirical findings. Likewise, empirical investigation cannot properly be pursued without conceptual clarification of what it is that is being investigated. So, for example, while philosophy can tell us how we should conceptualize self-consciousness, we need empirical studies to find out whether a particular organism meets the so-established criteria for self-consciousness or not, or how the abilities required for self-conscious thought are realized on a computational or neuronal level. Hence we need a collaborative approach if we want to know, for example, whether we share the ability for self-conscious thought with other animals, or whether self-consciousness is indeed a specifically human characteristic. On the other hand, empirical results that speak to this question may also, in turn, influence the way we conceptualize certain issues, as well as have an impact on ethical questions, such as how we should treat other animals. Thus, in keeping with this broadly naturalistic approach, the philosophical considerations that form the core of the book will be informed by empirical results, in particular from developmental psychology and ethology.

Third, since the notion of representation is an important concept for the book, let me explain how I understand this notion. Generally speaking, mental states, such as thoughts, desires, perceptions, beliefs, and so on, are states with intentional content, which is to say that they represent or are about things. Their content may be evaluated with respect to truth, accuracy, or appropriateness. In specifying the content of a mental state, we specify the way a state of affairs is presented to the organism, or, in other words, the way a state of affairs is taken to be by the organism. Mental representations are standardly analyzed in terms of an agent's intentional relations (e.g., propositional attitudes) toward (propositional) contents. So we might say that Peter hopes that the sun will be shining, or Susie knows that the leaves of trees are green, or Robin believes that there is a cat on the mat. Note that I take the expression "intentional relation" to be broader than the term "propositional attitude" because the former does not imply that the representational content of the mental state in question must be propositional. Since I will be investigating the notion of nonconceptual content in this book (which is arguably nonpropositional), I will therefore often use the term "intentional relation."

Representational states need to be distinguished from informational states. Although all representational states are informational, not all informational states are representational. For instance, although the rings in a tree trunk carry information about the age of the tree, and although smoke carries information about the likely presence of a fire, it would be wrong to say that tree trunks represent age or that smoke represents fire. This is, as has often been pointed out, because the reason that, say, a tree trunk carries information about the age of a tree lies in the fact that there is an invariant relation between the tree's trunk and its age. But this is not true for representational states. In contrast to purely informational states, representational states can *mis*represent.

We appeal to representational states when we want to explain the intentional behavior of an organism, that is, when we want to explain the organism's behavior in terms of the aims and desires that it is intended to satisfy while taking into account the organism's perceptual representation of the environment. According to the principle of parsimony, we should only do so if no simpler explanation is available for the same behavior (cf. Bermúdez, 1998, chap. 4). So, for instance, when a behavior can be explained in terms of simple stimulus-response reactions, no appeal to representational states of the organism need be made. I return to these issues in more detail in chapter 2.

I will now give a brief overview of the book's structure. Chapter 1 sets the stage for the discussion that follows. It spells out the problem of self-consciousness and locates it in its broader historical context. Readers who are familiar with the problem of self-consciousness will recognize most of the arguments discussed in this chapter, so they may wish to skip or skim over it. The chapter does not aim to produce novel insights, but it does provide the context necessary to fully appreciate the challenges faced by any theory of self-consciousness. Moreover, the chapter can be seen as an introduction (albeit brief) to the problem of self-consciousness for readers who are less familiar with the debate. In particular, the chapter presents traditional approaches to self-consciousness that conceived of self-consciousness as the result of a process of a subject reflecting on itself, and shows how these approaches lead into a regress. As a response to the regress problem, some authors have claimed that we need to posit the existence of a prereflective self-consciousness that can provide the grounding for higher-level, reflective self-consciousness. However, this raises the question of how this prereflective self-consciousness should be analyzed. I then analyze self-consciousness in terms of "I"-thoughts and investigate whether approaches that rely on an analysis of the semantics of the first-person pronoun can overcome the problems encountered by traditional models. I argue that while these linguistic approaches provide important insights with regard to crucial features of "I"-thoughts—such as their relevance for action and their immunity to error through misidentification—they ultimately remain incomplete, for they seem to presuppose the ability to nonaccidentally refer to oneself in thought, rather than providing an explanation for this possibility. This leads us to the question of whether recent appeals to nonconceptual forms of (self-)representation can provide a solution to this problem, as has been claimed by some contemporary authors.

The notion of nonconceptual content was originally introduced in the philosophy of perception as an alternative to traditional notions of representational content. Chapter 2 briefly analyzes the debate around nonconceptual content independent of the problem of self-consciousness—although with a focus on the arguments that are most relevant to the questions addressed in this book. Again, readers familiar with the debate around nonconceptual content will recognize most of the arguments presented here and may want to skim this chapter. However, it is important to have a proper grip on the notion of nonconceptual content to be able to evaluate theories of nonconceptual self-consciousness. I argue that there are indeed a number of good arguments that speak in favor of nonconceptual content. For instance, as we will see, beings that cannot be attributed with concept possession, such as animals and infants, do sometimes display intentional behavior that can only be explained if we credit these beings with representational states, though they lack conceptual abilities. In other words, they display primitive forms of normativity that warrant the ascription of nonconceptual representational states. Further, I argue that to determine the content of these states, we ought to appeal to the organism's skills; that is to say, nonconceptual content should be understood in terms of "knowledge-how" rather than "knowledge-that." This will have important implications for the analysis of implicit and explicit forms of representation presented in chapter 5.

Chapter 3 then takes us back to the topic of self-consciousness. The chapter discusses proposals that have been made for the existence of *nonconceptual self-consciousness* (see esp. Bermúdez, 1998). These proposals point to the fact that perception and bodily awareness are inherently self-representational, and argue that because of this they should be regarded as forms of nonconceptual self-consciousness. However, I argue that a critical flaw in such self-representationalist theories of nonconceptual self-consciousness consists in their neglect of the difference between implicitly self-related information and explicit self-representation. While perception and bodily awareness do in fact contain implicitly self-related information, they do not explicitly represent the self. Indeed, it would put an unnecessary and, for reasons of parsimony, implausible cognitive burden on the organism to explicitly represent the self in perception and bodily awareness. Thus self-representationalist theories misconstrue the nature of perception and bodily awareness. As a consequence, they fail both in giving an account of the possibility of "I"-thoughts (which require explicit reference to the self) and in giving an account of immunity to error through misidentification. Moreover, in their attempt to demonstrate that the self is represented in perception and bodily awareness, self-representationalist theories remain committed to the problematic subject–object model of self-consciousness, which I reject in chapter 1.

Accordingly, chapter 4 argues in favor of what I will call a non-self-representationalist theory (or "no-self" theory). According to this theory, the self is not part of the representational content of, for example, perception and bodily experience; rather, it is part of the *mode* of presentation. Consequently, such a theory avoids the objection of cognitively overburdening nonconceptual forms of representation, such as those involved in perception and bodily awareness. Moreover, it is able to provide a better account of immunity to error through misidentification. The theory defended here shares important similarities with those defended by Peacocke (1999), Recanati (2007), and O'Brien (2007). However, the theories defended by these authors remain incomplete, for they do not explain the transition from implicitly self-related information (where the self-relatedness is a function of the mode of experience) to explicit self-representation (where the role of the mode is made explicit via the application of the self-concept). That is to say, they leave open the question of how implicitly self-related information becomes conceptualized, and how a subject acquires a self-concept.

Chapter 5 takes the first step toward answering this question. Drawing on Karmiloff-Smith's (1996) theory of "representational redescription," I present an account of the distinction between implicit (i.e., procedural) and explicit representation and of the transitions between the two. In doing so, I argue in favor of a multilevel account of implicit and explicit representation that goes beyond the overly simplistic dichotomy suggested by the contrast between conceptual and nonconceptual representation.

Chapter 6 builds on the analysis of different levels of explicitness provided by chapter 5 and presents a multilevel account of the transition from implicitly self-related information to explicit self-representation. Here I argue that explicit self-representation develops over the course of an increasingly complex process of self–other differentiation that enables subjects to contrast their own mental and bodily states with those of others, thereby developing a concept of themselves as an individual entity that can be identified and referred to by others. As a result, we can distinguish between different levels or degrees of self-awareness. Further, we will see that self-consciousness cannot be understood without taking into account its constitutive relation to intersubjectivity. Both self-consciousness and intersubjectivity develop in parallel and are two sides of the same coin.

Chapter 7 spells out the implications of the theory presented here for the question of whether we share the ability to represent ourselves in thought with nonhuman animals. As we will see, the evidence to date is sketchy, but it does suggest that we can attribute some nonhuman animals (in particular chimpanzees) with some basic forms of self-awareness (and thus with what might be called a partial grasp of the self-concept).

Chapter 8 summarizes the results of the investigation, provides a conclusion, and presents open questions and avenues for future research.

# 1 Setting the Stage: The Problem of Self-Consciousness

## 1.1 Introduction

At this particular moment in time, I know that I am sitting at my desk, looking at the monitor in front of me, thinking about what to write. I am aware of my slight feeling of hunger and of the fact that my foot is itching. And I seem to be aware of all of this in a very direct, immediate way. In fact, it seems that I know these things in a special way, from the *first-person perspective*, or from the inside. To be sure, someone else could know these facts about me as well, but they would have to rely on my telling them or on observing my behavior. They could observe me scratching my foot and therefore conclude that it must have been itching, for example. But I do not need to rely on such observations to know of my feeling of hunger or that my foot is itching. I just know. It is this direct, unmediated awareness of one's mental and bodily states—whose canonical expression involves the first-person pronoun—that we are ultimately trying to understand when we discuss self-consciousness.[1]

Of course, there are many other things that I can know about myself but do not have access to in this immediate way. For instance, I know that I was born in January, because my parents told me so, and I can know that I am wearing a pair of blue socks, because I just looked down at my feet. I know what my face looks like from looking in the mirror or from looking at pictures of myself. The way I come to know about these things is not in principle different from the way someone else can come to know them. All these facts about myself can also be learned by someone else, in much the same way that I learn about them, namely, from the *third-person perspective*.

In fact, I can even come to know facts about myself from the third-person perspective without realizing that they are about myself. For instance, someone might tell me, "Kristina Musholt was born in January," and I could fail to understand that this is a piece of information about myself because I am suffering from amnesia and have forgotten my name. Or I might look at an old picture of me as a child and think, "What funny clothes that child is wearing!" without realizing that the child was myself. I might even, in passing, glance at a mirror and—without realizing that I am looking into a mirror—think, "That person really needs a haircut!" while failing to notice that the person is myself. This is not possible when I feel an itch in my foot and on the basis of this sensation come to think, "I have an itchy foot," or when I am feeling hungry and on the basis of this feeling think or say, "I am hungry." I know these things immediately, and there can be no doubt about who it is that is hungry or has an itchy foot when I form the corresponding judgment. I may have forgotten my name, and I may know nothing else about myself, but I can still refer to myself with the first-person pronoun to self-ascribe a state of hunger or a feeling of itchiness. Such cases of self-ascription are at the center of most theories of self-consciousness, as well as of this book. (Again, self-consciousness has to be distinguished from consciousness more generally. As mentioned in the introduction, consciousness more generally is not the topic of this book.)

As I just indicated, the canonical expression of the self-ascriptions in question involves the first-person pronoun, and they are such that their content can be known immediately and there can be no doubt as to who is their subject. Thus, as a first take on the phenomenon, we can define self-consciousness as the ability to think "I"-thoughts, that is, the ability to think thoughts that are about oneself and are known to be about oneself by the subject entertaining them. What exactly is the structure of this immediate self-knowledge—understood as the ability to refer to oneself in thought—and how do we acquire it? These are the central questions of the book.

This chapter sets the stage for an answer to these questions. It begins by presenting the traditional model of self-consciousness, according to which self-consciousness is a form of object cognition, and by explaining why this model is misguided (sec. 1.2). Based on this, I introduce the notion of prereflective self-consciousness and argue that this notion needs to be further analyzed (sec. 1.3). Section 1.4 considers linguistic approaches to the problem of self-consciousness and discusses the notion of immunity to error through misidentification, which I show to be an essential feature of paradigmatic forms of self-consciousness. Finally, section 1.5 argues that linguistic approaches remain incomplete, because they presuppose the ability to refer to oneself in thought without being able to explain how this ability can come about. This motivates the attempt to provide a theory of self-consciousness with the help of the notion of nonconceptual content.

## 1.2 The Traditional Subject–Object Model and Its Failure

One natural but misguided way of thinking about self-consciousness is to think of it in terms of the subject taking itself as an intentional object of perception or reflection. This way of thinking about self-consciousness goes back to what might be called the traditional model of epistemology. According to this model, which arguably dates back to Descartes, all knowledge is knowledge of objects. Accordingly, self-consciousness, too, must be explained in terms of knowledge of a particular type of object. So the model posits that in self-consciousness, the subject takes itself as the object of reflection or (inner) perception to gain self-consciousness (Frank, 1991a; Gloy, 1998).

This model turns out to be problematic for two reasons: First, it does not seem to do justice to the phenomenology of introspective experience, for as was famously pointed out by Hume, the self seems to escape any act of introspection. Second, as I have already mentioned, it either leads into an infinite regress or ends up being circular.

In Descartes's view, self-knowledge was constituted by a form of direct introspective access to the "thinking thing" that is our self. In contrast, Hume famously denied that we have introspective access to a self as such, for self-reflection always confronts us with a particular perception:

For my part, when I enter most intimately into what I call myself, I always stumble on some particular perception or other, of heat or cold, light or shade, love or hatred, pain or pleasure. I never catch myself at any time without a perception, and never can observe any thing but the perception. (Hume, 1967, p. 252)

So according to Hume, the phenomenology of introspection is such that it is devoid of any direct access to the self. We simply cannot perceive the self in introspection; it necessarily remains elusive. Accordingly, rather than conceiving of the self as an immaterial and unchangeable substance, Hume suggests that we should think of the self as "a bundle or collection of different perceptions" (ibid.).[2] The traditional model is therefore mistaken in trying to locate the self as such in the content of introspective experience.

But this model is also mistaken for another reason: we cannot explain how the self can be grasped, and, in other words, how self-consciousness is possible, as long as we conceive of self-consciousness as a relation between a subject and an object. This is because a model that conceives of self-consciousness in terms of a relation between a subject and an object leads into a regress. As has been pointed out by members of the "Heidelberg School" (Tugendhat, 1979)—comprising authors such as Dieter Henrich, Manfred Frank, and Ulrich Pothast—who in turn took their inspiration from Fichte, this is because for a subject to take itself as an object of its self-reflection, it needs to perform an identity judgment. In other words, to know that the object of my self-reflection is in fact myself, I need to self-identify. This requires that I recognize a certain property to be mine. But this act of self-recognition in turn relies on another act of self-identification, and so ad infinitum. Alternatively, to stop the regress, the model must assume that I already possess self-consciousness, in which case the model is circular (Henrich, 1967; Pothast, 1971; Frank, 1991b).[3]

Within the analytical tradition, Shoemaker has issued a similar criticism. Shoemaker targets the "inner-perception" model of self-consciousness in particular. Similar to the criticism brought forward by Fichte and the Heidelberg School, Shoemaker argues that self-consciousness cannot ultimately rely on self-perception, or in fact on any kind of subject–object relation, for to perceive myself as myself, I must already possess a first-personal knowledge of myself that can help me to identify a certain property as my own.

The latter point is especially important; it shows that the knowledge in question is radically different from perceptual knowledge. The reason one is not presented to oneself "as an object" in self-awareness, is, that self-awareness is not perceptual awareness, i.e. is not the sort of awareness, in which objects are presented. It is awareness of facts unmediated by awareness of objects. But it is worth noting that if one were aware of oneself as an object in such cases (as one is in fact aware of oneself as an object when one sees oneself in a mirror), this would not help to explain one's self-knowledge. For awareness, that the presented object was $\varphi$, would not tell one, that one was oneself $\varphi$, unless one had identified the object as oneself; and one could not do this unless one already had some self-knowledge, namely the knowledge, that one is the unique possessor of whatever set of properties of the presented object one took to show it to be oneself. Perceptual self-knowledge presupposes non-perceptual self-knowledge, so not all self-knowledge can be perceptual. (Shoemaker, 2003, p. 104)

Hence, to explain the possibility of self-knowledge, we need to assume the existence of a form of self-consciousness that is not to be understood as a relation between a subject and an object (which happens to be identical to the subject), whether in the form of a self-reflection or self-perception. Any model that conceives of self-consciousness as a form of object cognition will either lead into a regress or presuppose what it attempts to explain.

It is all the more revealing that, as we will see later, some contemporary theories of nonconceptual self-consciousness—namely, those that I will call "self-representationalist theories"—remain implicitly committed to the subject–object model of self-consciousness, which renders them vulnerable to precisely the two kinds of mistake that I just pointed out.

## 1.3   Prereflective Self-Consciousness

As a consequence of the failure of the subject–object model, representatives of the Heidelberg School argued that self-consciousness cannot possibly be described as a relation, that is, as something reflecting on itself or as something perceiving itself. In their account, every possible self-ascription of any property necessarily presupposes the possession of a more fundamental form of self-consciousness, that is, a kind of immediate familiarity with oneself, or prereflective self-consciousness. This echoes Fichte's thought that the self must possess an immediate acquaintance with itself, which he described as a complete nondifferentiability of subject and object in self-consciousness (Frank, 1991a). As we have just seen, without this prereflective, nonrelational, nonidentificational form of self-consciousness, I could not possibly know that it is *I* who possesses the property in question.

The call for a prereflective self-consciousness also reverberates through much of the phenomenological literature (Zahavi, 2005).[4] For instance, according to Zahavi, Merleau-Ponty (1962) argues that consciousness is always "given to itself," and Sartre (1966) argues that every experience comes with a prereflective self-consciousness. In fact, Sartre seems to claim that prereflective self-consciousness is the only mode of existence by which it is possible to be conscious of anything (Sartre, 1966, p. 20). In other words, not only is prereflective self-consciousness necessary to explain the possibility of reflective self-consciousness on this view, but it is also a necessary structural feature of all conscious experience. The notion of a prereflective self-consciousness can also be found in Heidegger's work (see Zahavi, 2005, chap. 1). This prereflective self-consciousness is often said to be based on a particular kind of first-personal "givenness" or a sense of "mineness," that is, "the fact that the experiences are characterized by a first-personal givenness that immediately reveals them as one's own" (Zahavi, 2005, p. 124).[5]

But while the postulation of a prereflective self-consciousness seems to solve the problem posed by the traditional, reflection-based model of self-consciousness, as it stands, this solution remains dissatisfying. It remains unclear how this prereflective self-consciousness is to be understood. In fact, some authors go so far as to explicitly state that prereflective self-consciousness cannot be analyzed further and has to be taken as a fundamental "given": "However, we must also humbly declare that the basic element of our theory, familiarity, cannot be further analyzed" (Frank, 2002, p. 400).

Thus, it seems, we are left with a purely negative characterization of the phenomenon. We have learned that self-consciousness is not to be understood as a form of perception or reflection that takes the self (or one of its mental states) as its object, for this way of conceptualizing the phenomenon is circular. Self-consciousness is not like object awareness; rather, the self is aware of itself "as a subject." It is not mediated; rather, it is immediate and nonobservational. But we have yet to learn how to make sense of self-consciousness in terms of a positive characterization.[6] In other words, to say that self-consciousness must be based on some kind of prereflective, immediate familiarity with oneself is to describe the problem, but it does not yet help us understand how to solve it.

In the following chapters, I want to challenge Frank's claim "that the basic element of our theory, familiarity, cannot be further analyzed" (Frank, 2002, p. 400). Indeed, I take it that theories of nonconceptual self-consciousness can in a sense be seen as picking up where theories of prereflective self-consciousness leave off. However, as mentioned earlier, what I call "self-representationalist theories" remain implicitly committed to the subject–object model of self-consciousness. Hence they cannot provide us with a satisfying analysis of prereflective self-consciousness. Instead we should aim for what I call a "non-self-representationalist account." As we will see in chapter 4, what is crucial according to such an account is the distinction between the representational *content* and the *mode* of an experience. In this view, the self is not part of the representational *content* of experience; rather, it is part of the *mode*. It follows that the "sense of mineness" is nothing above and beyond certain functional features of conscious mental states. Nonetheless I take this to be compatible with a certain reading of phenomenological approaches, as well as approaches in the tradition of German idealism. In this reading, these approaches argue precisely that in prereflective self-consciousness the self-involving nature of conscious experience is in some sense phenomenologically available without the self being represented in experience.[7]

## 1.4 Analyzing the Use of "I"

Before we turn to theories of nonconceptual self-consciousness, it is first necessary to consider linguistic theories of self-consciousness. It has been suggested that a turn to language might help us make progress in solving the problem that traditional reflection-based models of self-consciousness have left us with without requiring us to assume the existence of a prereflective sense of self (e.g., Tugendhat, 1979). The rationale for this turn is that one way (and perhaps the best or, according to some authors, even the only way) to analyze a thought is by analyzing its canonical linguistic expression. As Dummett puts it: "It is the essence of thought not merely to be communicable, but to be communicable, without residue, by means of language. In order to understand thought, it is necessary, therefore, to understand the means by which thought is expressed" (Dummett, 1978, p. 442). As we have seen, the canonical linguistic expression of self-conscious thoughts involves the first-person pronoun. Thus, according to Dummett's rationale, to understand the structure of self-consciousness and its underlying representational abilities, we should investigate how the first-person pronoun is used. Perhaps we can, as Tugendhat (1979) has argued, avoid the problems associated with the subject–object model of self-consciousness by turning our attention to the use of the first-person pronoun.

Let us consider the semantics of the first-person pronoun, then. The semantic role of the first-person pronoun is such that it necessarily refers to the utterer of a sentence (or the thinker of a thought) containing it. In other words, it has guaranteed self-reference. According to Tugendhat (1979), to ascribe predicates to oneself by means of the first-person pronoun, a speaker does not need to know anything about herself as a person; she does not need to know any self-identifying criteria. That is to say, these self-ascriptions are identification free and do not rely on self-observation, whereas other-ascriptions do rely on observation (see also Evans, 1982;

Shoemaker, 1968). Thus there is, in Tugendhat's terminology, an "epistemic asymmetry" between self-ascriptions of mental predicates and other-ascriptions of mental predicates. A thinker or speaker who self-ascribes a mental state by means of the first-person pronoun does not have to self-identify but only has to know that the person she refers to by means of the first-person pronoun is, in principle, *identifiable* by others by means of singular terms like names or definite descriptions, or by means of the third-person pronoun—just as she can identify others and refer to them by means of singular terms, while these others, in turn, can also refer to themselves by means of the first-person pronoun.[8] Thus a first-personal self-ascription of a predicate has the same truth conditions as a third-personal ascription of that same predicate to the same person made by someone else (cf. Rosefeldt, 2000). So my utterance "I have a headache" has the same truth conditions as the utterance "She has a headache" or "The person over there has a headache" or "Kristina has a headache" made by someone else (provided that that someone refers to me). Tugendhat calls this the "principle of veritative symmetry."

### 1.4.1  "I"-Thoughts and Their Relation to Action

It is important to see that despite this "veritative symmetry," the first-person pronoun in self-ascriptions cannot be reduced to or replaced by a name or definite description. This is because self-ascriptions that involve the first-person pronoun may have the same truth conditions as third-person ascriptions, but they do not have the same cognitive significance for their subject (cf. Rosefeldt, 2000). Only self-ascriptions involving the first-person pronoun (e.g., "I have a headache"), but not ascriptions involving a name or a definite description (e.g., "KM has a headache"), have immediate implications for action.[9] Consider this well-known story by John Perry:

I once followed a trail of sugar on a supermarket floor, pushing my cart down the aisle on one side of a tall counter and back the aisle on the other, seeking the shopper with the torn sack to tell him he was making a mess. With each trip around the counter, the trail became thicker. But I seemed unable to catch up. Finally it dawned on me. I was the shopper I was trying to catch. (Perry, 2000, p. 27)

What is happening here? Obviously Perry was aware from the start that *someone* was making a mess, and this thought did *de facto* refer to him. Nevertheless he was not aware that the thought referred to him until he realized that *he himself* was making the mess. We have already seen that we sometimes have information about ourselves without realizing that it is about ourselves. Perry's example is one of these cases. What Perry is after here is the fact that one can entertain many thoughts that are *in fact* about oneself, such as, in his example, "Someone is making a mess," "The only bearded philosopher in the market is making a mess," or even "John Perry is making a mess," without realizing that these thoughts refer to oneself; one might suffer from amnesia and hence might not remember one's name, for example, or fail to realize that the description "the only bearded philosopher in the market" refers to oneself. In contrast to those kinds of beliefs about oneself, there is a class of thoughts, namely, *de se* thoughts, that can only be entertained by a subject if that subject is aware that the thoughts are self-referring. In other words, these thoughts are nonaccidentally self-referring. The canonical expression of these thoughts involves the first-person pronoun. Perry can only come to think "I am making a mess!" if he realizes that it is in fact he himself who is carrying a leaking sugar bag, for to competently employ the first-person pronoun, he has to know that it necessarily refers to him when he is entertaining this thought. And it is this nonaccidentally self-referring thought that will prompt him to stop the cart and fix the leak.

*De se* beliefs have to be distinguished from *de re* or *de dicto* beliefs. An example of a belief *de dicto* is a classical propositional attitude ascription of the form "Max believes that Lisa is going to Spain next week."[10] In this example, Max believes the *dictum* (what is being said) or proposition, namely, "that Lisa will go to Spain next week." But we might also say that he has a belief about someone (or something), namely, Lisa, which is to say that he has a belief *de re*. This can be expressed by saying, "Max believes *of* Lisa that she is going to Spain next week." Moreover, Lisa can also have a belief about herself, which, as we have just seen, if she also knows that it is about herself, we can call a belief *de se*. We might say, for instance, "Lisa beliefs that she* (herself) is going to Spain next week."[11] But, as we have just seen in the Perry example, not all beliefs about oneself have to be *de se*. For example, Lisa might have a *de dicto* belief about herself, such as "Lisa believes that the best student in the philosophy department will go to Spain next week," where the best student in the philosophy department is identical to her. In this case, she has a belief about herself, but it is possible that she fails to realize that this belief is actually about herself, if, for example, she doesn't know or believe that she is the best student in the philosophy department. She might also have a *de re* belief about herself, which we could express as "Lisa, who is identical with the best student in the philosophy department, believes of the best student in the philosophy department that she is going to go to Spain next week." This clearly is a belief that Lisa has about herself, but it does not imply that she is also aware that this is a belief she has about herself. As we have just seen, Lisa might be identical with the best philosophy student without realizing it. Or she might identify the person going to Spain next week without realizing that it is herself. In other words, beliefs *de dicto* or *de re* do not imply beliefs *de se*, thus suggesting that first-person thoughts are irreducible to thoughts *de dicto* or *de re* (Castañeda, 1966; Kaplan, 1977; Lewis, 1979; Perry, 1979). What this shows is that one can have beliefs that are in fact referring to oneself without being aware of this fact.

The latter, namely, awareness of the self-referring nature of a belief, is required for that belief to have guaranteed implications for action. It is only by realizing that a thought is about myself that I can take appropriate action (such as, in Perry's example, stopping my cart to rearrange the sugar bag). So in contrast to *de re* or *de dicto* thoughts about myself, *de se* thoughts, that is, thoughts that are known by me to be about myself, have immediate relevance for action. Again, these thoughts are specified in direct speech by means of the first-person pronoun (and, following Castañeda, in indirect speech by means of she* or he*). In other words, when Lisa is thinking of herself, "I am going to Spain next week," or when I am thinking, "I am making a mess," there can be no doubt as to whom these thoughts are referring to, and hence these thoughts are guaranteed to influence the subject's action; they might, for instance, prompt me to stop making a mess, and they might prompt Lisa to begin to prepare for her trip. This is in contrast to *de re* or *de dicto* thoughts about oneself that do not involve the first-person pronoun. Therefore the first-person pronoun is also referred to as the "essential indexical" in this context (Perry, 1979).

### 1.4.2  Immunity to Error through Misidentification

However, as Wittgenstein (1958) pointed out, on closer examination, we can actually distinguish between two different uses of the first-person pronoun: the use "as subject" and the use "as object."[12] Examples of the use "as object" are thoughts such as "I need a haircut" or "I have grown six centimeters." Examples of the use "as subject," are "I see a cup of coffee in front of me" or "I hear the birds outside my window." To understand what Wittgenstein is after with this distinction, it is helpful to think about what kind of questions you can reasonably ask a speaker who uses the first-person pronoun. If the speaker uses the pronoun as object, as in "I have grown six centimeters," it is theoretically possible that she is mistaken with regard to the question of *who* has grown six centimeters. For example, you could think of a situation in which the subject, upon looking at some markings that her parents made on the wall, forms the thought that she has grown six centimeters based on the fact that she mistook the markings as representing her own height, though they in fact represent the height of the subject's sister. In such a situation, you would have reason to ask, "Are you sure that it is you who has grown six centimeters?" and the subject would in fact commit an error of misidentification when claiming that it is she herself who has grown. Contrary to this scenario, the speaker can principally not be mistaken with regard to the identity of the subject when she says, "I see a cup in front of me." In this case, the subject could be mistaken about the fact that it is a cup that she sees (perhaps mistaking it for a different object), but the subject cannot possibly be mistaken about the fact that it is *she* who has this (possibly nonveridical) perception. Accordingly, it would make no sense to ask, "Are you sure that it is *you* who is seeing the cup?" In this latter case, the subject cannot commit an error of misidentification relative to the first-person pronoun (Shoemaker, 1968). Shoemaker introduces the notion of immunity to error through misidentification in the following way:

To say that a statement "a is φ" is subject to error through misidentification relative to the term "a" means that the following is possible: the speaker knows some particular thing to be φ, but makes the mistake of asserting "a is φ" because, and only because, he mistakenly thinks that the thing he knows to be φ is what "a" refers to. (Shoemaker, 1968, p. 557)

The kind of mistake Shoemaker describes is not possible in statements that are immune to error through misidentification with respect to the term "a." And this is precisely the case in those statements (and, accordingly, the thoughts that are expressed by those statements) in which the first-person pronoun is used "as subject." So while the thought "I have grown six centimeters" in the earlier example *is* susceptible to error through misidentification with respect to the first-person pronoun, it seems that thoughts such as "I am feeling angry" or "I am in pain" are immune to such error.

In other words, if I were to think, and assert, "I have grown six centi-meters," the question "Are you sure it is you who has grown?" would be a sensible one. The question is sensible because my basis for forming the judgment does not supply sufficient warrant for the self-attribution con-tained in the judgment. On the other hand, when I feel anger and on the basis of this form the judgment "I am angry," my being asked, "Are you sure that it's you who is angry?" "would be nonsensical" (Wittgenstein, 1958, p. 67). It is simply not possible to be feeling anger only to discover that the subject of this experience is not me. The question is nonsensical because my basis for forming the judgment—my experience of being angry—pro-vides sufficient warrant for the self-attribution involved in the judgment.

Of course, when thinking, "I am feeling anger," I might be wrong as to whether it is anger that I am feeling; perhaps the feeling would more accurately be described as an irritation (or hunger). However, the important point here is that, in having this thought, I *cannot* be mistaken about who it is that is allegedly feeling anger. So a thought's being immune to mis-identification allows for the possibility that it misrepresents the property that is being ascribed, and hence should not be confused with infallibility. However, it cannot misrepresent the subject purportedly possessing that property. In other words, such a thought can "what"-misrepresent, but it cannot "who"-misrepresent (Meeks, 2006).

The reason for this, according to Wittgenstein, Shoemaker and Evans, is that these self-ascriptions do not involve the identification of a subject. As Evans (1982) puts it, such judgments do not contain an identification component. If no identification of a subject is involved, it is impossible to misidentify it. This is why Wittgenstein terms the use of the first-person pronoun involved in these kinds of first-person judgments the use "as sub-ject" as opposed to the use "as object." When making a first-person judg-ment that is immune to error through misidentification, we are expressing a type of awareness of ourselves not as objects that have to be identified but as subjects of thought or experience.

Now, according to Wittgenstein, the immunity principle only applies to the self-ascription of mental states, not to the self-ascription of bodily states.[13] Accordingly, Wittgenstein argues, it is quite possible for me to be mistaken about whether a broken arm is mine, but it is not possible to be mistaken about whether a pain that I experience is my own. In contrast, Evans (1982) holds that immunity to misidentification can apply to bodily self-ascriptions as well, given that they are based on the right kind of evidence (i.e., evidence "from the inside," through somatic proprioception). He points out that "it seems equally not to make sense for a subject to utter 'someone's legs are crossed, but is it I whose legs are crossed?,' when the first component is expressive of knowledge which the subject has gained about the position of his limbs, available to him in the normal way" (Evans, 1982, p. 216). As we will see in more detail in chapters 3 and 4, it makes sense to side with Evans on this matter.[14] However, for the moment, it is only important to understand the claim that there are *some* self-ascriptions that are immune to error through misidentification, which is to say that they are made based on certain grounds. These grounds are such that they are necessarily self-involving, in a way that is to be further specified. Hence they do not require any kind of self-identification.

Why is this so important if, as we have seen, it is not true for all thoughts involving the first-person pronoun? As we have seen, immunity to misidentification is not a feature of thoughts that involve the use of the first-person pronoun "as object." In contrast to thoughts that involve the use of the first-person pronoun "as subject," thoughts that rely on an objective use of the first-person pronoun involve an identification component, which is why they allow for the possibility of misidentification. Nonetheless these thoughts ultimately have to be grounded in thoughts that are immune to misidentification (cf. Bermúdez, 1998, p. 7). As was explained earlier, necessarily, every self-identification already presupposes some kind of self-knowledge: "Where the use of 'I' *does* involve an identification, the making of the identification will always presuppose the prior possession of other first-person information" (Shoemaker, 1996, p. 211). Thus, to even get self-identification off the ground, we have to assume that there is some kind of first-person information that does not rely on self-identification. In other words, for there to be any possibility of self-consciousness at all, there has to be a form of self-consciousness that is not based on identification, which is to say that there has to be a type of first-person judgment that is immune to error through misidentification. Hence "I"-thoughts that are immune to error through misidentification (i.e., where the "I" is used "as subject") are epistemologically prior to those that are not immune to error through misidentification (i.e., where the "I" is used "as object").

Thus, in a sense, the analysis of first-person thoughts in terms of immunity to error through misidentification is just another way of arriving at the points made by Fichte, the Heidelberg School, and certain phenomenologists, namely, that self-consciousness is not ultimately to be understood as an awareness of oneself as an object, and is not to be understood in terms of perception or some other type of self-reflection on pain of regress. This is why immunity to error through misidentification is an essential feature when we are trying to understand self-consciousness, even if it does not apply to all self-conscious thoughts. Although not all self-conscious thoughts are immune to misidentification (those that involve the use of "I" "as object" and rely on an identification component are not), there must be *some* self-conscious thoughts that are immune if we are to explain the possibility of self-conscious thoughts at all.

Immunity to misidentification is also importantly related to the feature of "I"-thoughts mentioned in the previous section, namely, their immediate relevance for action. Because self-conscious thoughts that are immune to misidentification do not rely on an identification of the subject, there can be no room for the question of who the subject of the thought is. When I experience a headache and seek relief by taking an aspirin, I do not first have to identify the subject of the pain experience to seek ways to ameliorate the pain. Rather, I can act immediately. I do not have to identify the subject of the pain experience because I cannot come to know about anybody else's pain in this same way (i.e., from the first-person perspective). In contrast, there will always be a potential gap between thought and action when a self-ascription has to rely on an identification component (cf. Bermúdez, 1998, ch. 1). For instance, if I see myself in a mirror having a stain on my face, I might fail to act to remove the stain if I fail to realize that the person I am looking at is myself. This is a possibility because I could potentially come to know about someone else having a stain on her face in the same way, that is to say, by observation (i.e., from the third-person perspective).[15] Similarly, Perry failed to stop his cart when leaking sugar because although he had information about himself—someone was leaking sugar, and that someone happened to be identical to him—this information was such that it could have been about someone else. Of course, once I form the relevant "I"-thought ("I have a stain on my face," "I am making a mess"), I will take action, but to be able to form this thought, I must first have some way of gaining self-knowledge that does not rely on self-identification and can provide the basis for those "I"-thoughts that do rely on self-identification.

Because of the fundamental role they play for understanding self-consciousness, "I"-thoughts that are immune to error through misidentification are the focus of the following investigation, even though, as we have seen, not all "I"-thoughts are immune to error through misidentification.

### 1.4.3 Anscombe's "No-Reference" View of "I"

Based on the insight that cases of self-ascription where the "I" is used "as subject" are immune to error through misidentification, Wittgenstein (1958) famously argued that in these cases the first-person pronoun does not refer at all. In this view, the utterance "I am in pain," for example, does not refer to the subject of the utterance, in the sense that it does not pick out a particular subject; rather, the utterance merely serves as an expression of the pain state. If the function of the first-person pronoun in such cases is not to pick out or to identify a particular person, then it cannot misidentify. Anscombe (1994) radicalized this thought and claimed that, in fact, the first-person pronoun *never* refers.

However, this claim is hard to swallow. After all, we do seem to make reference to ourselves when we use the first-person pronoun. Indeed, there are two ways of resisting Anscombe's claim. One the one hand, as Coliva (2012) observes, one might point out, as numerous authors have (e.g., Shoemaker, 1968; Evans, 1982; McDowell, 1998; Wright, 1998), that indexical and demonstrative judgments, such as "It is sunny here" or "That tree is green" also cannot misidentify, yet we do not think that because of this they fail to refer to a place or object. Thus, that a term cannot misidentify its referent in and of itself is not a reason to believe that the term does not refer at all.

On the other hand, O'Brien (2007) has provided an illuminating analysis of Anscombe's thesis, which ties it to an implicit adherence to the problematic traditional subject–object model of self-consciousness that we discussed earlier. In a nutshell, in O'Brien's view, Anscombe's thesis is a result of her belief that if the first-person pronoun is to refer at all, then its referent must be "really or physically" presented to the subject. In other words—having ruled out the possibility that the first-person pronoun functions like a proper name—she construes the reference of "I" along the lines of the reference of demonstratives by means of which we can refer to the objects we are presented with in perception. However, Anscombe argues, we can imagine a subject being put in a sensory deprivation tank, which would deprive the subject of any kind of bodily or external information. Nonetheless the subject would still be able to refer to herself in thought, for instance, by forming the thought "I won't let this happen again!" Given that, according to the view apparently assumed by Anscombe, to be able to self-refer in thought, the subject must be presented to herself in experience, and given that when in a sensory deprivation tank, the subject cannot be presented with her body, the first-person pronoun cannot refer to the body. Rather, the only way in which the subject can be presented to herself in this situation, and hence the only object that the first-person pronoun can refer to, is a kind of Cartesian ego, that is, a "thinking thing." But this, in Anscombe's view, would be an unacceptable consequence; hence we have to conclude that the first-person pronoun does not refer at all. Thus, in Anscombe's view, to escape the Cartesian conception of the self as being essentially a bodiless ego, a mere "thinking thing," we ought to assume that the first-person pronoun does not refer. However, as O'Brien rightly points out, we are only forced into this conclusion if we adopt the traditional subject–object model of self-consciousness in the first place. And we have already seen good reasons to resist this model. Accordingly, the lesson to be learned from this is that we should not construe self-reference along the lines of object reference to begin with—in line with the arguments that speak against the subject–object model discussed earlier. In other words, as O'Brien sees it, Anscombe fails to see that we need not construe self-consciousness along the lines of the subject–object model, which results in her defending the highly counterintuitive and problematic view that the first-person pronoun cannot refer. Rather than accepting Anscombe's conclusion, we should see her view as providing us with yet another reason to reject the view that self-consciousness is analogous to object awareness. So the ability to think "I"-thoughts is not to be seen as a result of the self being presented to itself in experience. Indeed, as we will see in chapters 3 and 4, "I"-thoughts possess the features that they do precisely because they express a form of awareness that *does not* represent the self. That is to say, Anscombe is right when she argues that what she calls "unmediated agent-or-patient conceptions of actions, happenings, and states" are "subjectless" (Anscombe, 1994, p. 159), though she is wrong to think that this means that the first-person pronoun does not refer.

### 1.4.4   A Threat to the Immunity Principle? The Case of Schizophrenia
As we have just seen, what makes certain first-person judgments immune to misidentification is the evidence on which they are based. Although it is

a matter of controversy as to what exactly the right kind of evidence bases are, again, there ought to be at least some kinds of evidence that provide a basis for first-person judgments that are immune to error through misidentification, for otherwise we would end up with a regress. Mental states such as being in pain or entertaining a thought are traditionally regarded as being among the strongest candidates for these. (Although, as we will see in chapters 3 and 4, some bodily forms of self-awareness are also immune to misidentification.) One potential counterexample against the existence of such evidence bases is the phenomenon of thought insertion that sometimes occurs in patients suffering from schizophrenia. This phenomenon manifests itself in certain reports made by patients, such as in the following example: "Thoughts come into my head like 'Kill God.' It's just like my mind working, but it isn't. They come from this chap, Chris. They're his thoughts" (Frith, 1992, p. 66). John Campbell (1999) takes this to be a counterexample to the immunity principle. Thought insertion is interesting because it seems to suggest that I can entertain a thought (and be aware of the thought) and yet fail to ascribe the thought to myself. This is problematic for the view defended here, for it seems as though the judgment in question is based on information that should be immune to misidentification—after all, one cannot become aware of anyone else's thought but one's own in this way—yet the judgment fails to self-refer. One could take this to suggest that what was taken to be a form of identification-free, nonobservational self-consciousness is based on identification after all, which would refute the claim that some forms of self-consciousness are immune to misidentification because their underlying information base is such that it cannot deliver information about anyone but oneself. Obviously, this would also reinvoke the regress worry associated with observation- and reflection-based models of self-consciousness and would again raise the question of how to provide a noncircular account of self-consciousness.

Of course, whether Campbell's claim can withstand scrutiny will depend in part on how to adequately capture the phenomenology of thought insertion. It is not always clear how literally one ought to take the patients' reports about these matters (after all, these patients do suffer from a psychopathology). But—for the sake of argument—let us take reports like the one cited earlier at face value and grant that the patient experiences awareness of a thought that she fails to ascribe to herself. Does this establish a counterexample to the immunity principle?

Some authors (e.g., Gallagher, 2000; Coliva, 2002) have argued that it does not, on the grounds that we ought to distinguish between the sense of ownership and the sense of agency of a thought, and in the case of thought insertion, only the sense of agency is affected, while the sense of ownership is left intact.[16] (Indeed, the distinction between the ownership and agency of a thought was already anticipated in Campbell's essay.) These authors argue that the thought is still experienced as the subject's thought, in the sense of being experienced in the subject's mind, although the subject lacks the sense of being the agent of the thought in question. After all, the subject is not confused with regard to *where* the thought occurs. The thought is experienced as occurring in her own mind: that is what makes the experience so terrible and leads the subject to complain about it. However, the subject lacks a sense of agency over the thought, that is to say, a sense of being the initiator or source of the thought. This is what causes her to attribute the thought's origin to someone else.[17] According to Frith (1992), the underlying mechanism that leads to the absence of a sense of agency is based on a breakdown of self-monitoring. Frith advocates a comparator model of thoughts that is analogous to the comparator model of motor actions. According to the classical comparator model of motor actions, when a motor instruction to move is sent to a set of muscles, a copy of that instruction, the efferent copy, is also sent to a comparator (Held, 1961; von Holst & Mittelstaedt, 1950). The comparator stores the efferent copy and then compares it to the reafferent visual or proprioceptive information about the movement that was actually executed. The lack of a sense of agency, according to this model, is due to a mismatch between efferent copy and reafferent information. In addition, the comparator model also contains a forward component of motor control. This allows it to register the efferent copy as correctly or incorrectly matching motor intentions, allowing for automatic corrections to movement before any sensory feedback is received (Georgieff & Jeannerod, 1998). Frith postulates a similar mechanism for thoughts and inner speech. He suggests that this mechanism is generally accompanied by a sense of effort, and defects in the mechanism can explain the phenomenon of thought insertion in schizophrenia:

Thinking, like all our actions, is normally accompanied by a sense of effort and deliberate choice as we move from one thought to the next. If we found ourselves thinking without any awareness of the sense of effort that reflects central monitoring, we might well experience these thoughts as alien and, thus, being inserted into our minds. (Frith, 1992, p. 81)

Frith's model has been criticized for a variety of reasons (see, e.g., Stephens & Graham, 2000; Gallagher, 2004; Jeannerod & Pacherie, 2004; Synofzik et al., 2008), and a variety of alternative models have been proposed (e.g., Gallagher, 2004; Synofzik et al., 2008; see also G. Carruthers, 2008). However, the details of this debate need not concern us here. While the question of exactly what underlying subpersonal mechanisms cause a lack of the sense of agency in cases of thought insertion is interesting, for the present purposes, the important point is that at the personal level, thought insertion does not seem to entail a lack of the sense of *ownership*. Although the subjects who are reported to experience cases of thought insertion do not experience themselves as having initiated the "inserted" thoughts, and thus lack a sense of agency, these thoughts are nonetheless experienced as occurring within the subjects' own minds. And insofar as "inserted thoughts" are still experienced as occurring within the subjects' own minds, the immunity principle has not been violated.[18]

But there is a more general point to be made with regard to the claim that thought insertion provides a counterexample to the immunity principle.[19] This point holds independently from whether or not one accepts the distinction between ownership and agency for thoughts.[20] The point is that the immunity principle claims that a special kind of informational link exists between some first-person judgments and their underlying evidence base. But this informational link guarantees not *that* one will form an appropriate first-person thought, but that *if* one forms a first-person thought (i.e., a thought that self-refers), there is no question as to who the subject of the thought is. This is because, as we saw earlier, the basis for forming the first-person thought is necessarily self-involving without requiring self-identification. When one forms a thought that someone else is having a thought of which one is aware, as is the case in thought insertion, one is failing to apply the first-person concept (hence one's thought fails to self-refer), but that does not violate the conditional mentioned earlier. Having the thought itself guarantees that it is *one's own* thought, but one can still fail to make the transition to an explicit self-ascription of the thought. So the point about immunity to misidentification is that the information base in question is essentially self-involving in the sense that *if* one forms a first-person thought based on this information, then there can be no further question as to who the subject of that thought is. And it is not an objection to this claim to say that it is possible to fail to form the first-person judgment in the first place. Various things could cause such a failure. For instance, if a subject does not have mastery of the first-person concept, she will be unable to apply it correctly. And even if the subject does possess the relevant concept, there are circumstances in which she might fail to apply it—as is arguably the case in thought insertion. Of course, it is an interesting question as to what exactly the causes of failure in such circumstances are. However, for present purposes, it is sufficient to understand that talk about immunity is just meant to indicate those mental states where *if* one has the first-person concept, and *if* one applies it on the basis of being in the mental state in question, then there cannot be an error through misidentification of the subject.[21] In other words, introspective awareness of a thought is sufficient to make it your own—and would therefore make a self-ascription on the basis of such introspective awareness rational—even if it does not guarantee that you will in fact ascribe that thought to yourself. Thus the phenomenon of thought insertion does not pose a counterexample to the immunity principle, even if we take the subjects' reports at face value, and even if we do not accept the distinction between ownership and agency as being sufficient to refute Campbell's claim.[22]

## 1.5   The Problem of Self-Consciousness Returns

Let us take stock. We started with the observation that we seem to be aware of (some) of our mental and bodily states from the first-person perspective, which is to say that we seem to have direct, immediate access to these states. In asking how to understand the structure of this self-awareness, we came to realize that it is not to be understood as a form of self-reflection or self-perception, as is assumed by traditional models of self-consciousness. However, the suggestion made by some authors that we must therefore assume the existence of a prereflective form of self-consciousness raised the question of how this should be analyzed (I address this question in chap. 4). We then turned to an analysis of the canonical linguistic expression of self-conscious thoughts by examining the use and function of the first-person pronoun. It turned out that the first-person pronoun, which is a characteristic component of canonical expressions of attitudes *de se*, cannot be reduced to definite descriptions, names, or other singular expressions, and thus attitudes *de se* cannot be reduced to attitudes *de dicto* or *de re*. We also saw that in contrast to attitudes *de dicto* or *de re*, attitudes *de se* are guaranteed to have immediate implications for action. In addition to this property, first-person judgments have the property of being immune to error through misidentification relative to the first-person pronoun, given that they are made based on the right kind of evidence base. This is because these judgments do not rely on self-identification, in turn confirming that self-consciousness is not to be understood as a form of self-perception or self-reflection. As a result of this analysis, we can state that self-consciousness can be defined as the ability to think "I"-thoughts (or *de se* thoughts)—where the "I" is used as subject—and these thoughts are characterized by nonaccidental self-reference, by the property of having immediate implications for action, and by being immune to error through misidentification.

So does this solve our problem of providing a noncircular account of self-consciousness that helps us understand the structure and possibility of entertaining "I"-thoughts? Despite the progress we have made in understanding the structure of "I"-thoughts, this is not the case. We still do not know how it is possible for a creature to entertain "I"-thoughts. In fact, it has even been argued that trying to solve the problem of self-consciousness by means of an analysis of the linguistics of the first-person pronoun leads us straight into a paradox (Bermúdez, 1998). This paradox arises from two core assumptions: First, to understand self-consciousness, you have to understand the capacity to think "I"-thoughts (where "I"-thoughts are thoughts that nonaccidentally refer to oneself, have immediate implications for action, and are immune to error through misidentification). Second, to understand the capacity to think "I"-thoughts, you have to analyze their canonical linguistic expression, that is, the use of the first-person pronoun (Bermúdez calls this the thought-language principle). The problem with these assumptions is that the ability to use the first-person pronoun seems to presuppose the ability to think "I"-thoughts. After all, as we have seen, the semantic role of the first-person pronoun is such that it necessarily refers to the utterer of the sentence containing it. So a speaker who is using the first-person pronoun correctly has to understand that it refers to himself (and, according to Tugendhat, that others can refer to himself by means of singular terms). But this understanding—"When I use the first-person pronoun, it always refers to myself"—is itself an "I"-thought. Accordingly, it seems that one cannot give an account of the mastery of the first-person pronoun without referring to the capacity to think "I"-thoughts. Thus we seem to have an explanatory circle, for any analysis of the use of the first-person pronoun must presuppose what it is trying to explain, namely, the ability to nonaccidentally refer to oneself in thought.[23] A second type of circularity arises because, given the interdependence of the capacity for self-conscious thought and mastery of the first-person pronoun, it is impossible to explain how self-consciousness can develop, either ontogenetically or phylogenetically. Children are not born with the ability to use the first-person pronoun, nor do we find this ability in nonhuman animals. However, as Bermúdez points out, for any ability that is psychologically real, we should be able to provide an account of how it can be acquired. One might argue that it is not the task of philosophy to provide such an explanation, for philosophy is concerned with conceptual analysis and not with providing developmental accounts. Nonetheless, as I argued in the introduction, philosophical theories should be empirically plausible in the sense that although they do not necessarily have to provide a developmental account of the phenomena they deal with, they should at least not exclude the possibility of providing such an account, and they should be compatible with the empirical data that speak to such an account. Hence we seem to have two forms of vicious circularity, an explanatory circularity and a capacity circularity.

Even if one does not agree with the claim that an analysis of self-consciousness in terms of the semantics of the first-person pronoun leads into an explanatory circle—for one might argue that the ability to think "I"-thoughts just *consists in* the ability to employ the first-person pronoun in language and thought and hence cannot be circular with the latter[24]—one would still want to know what enables a creature to think of itself in a way that allows it to refer to itself by means of the first-person pronoun. Put differently, we need an independent account of first-personal ways of thinking. That is to say, we need answers to questions such as "What are the underlying representational abilities that enable a creature to have 'I'-thoughts?"; "Why do these thoughts have the specific characteristics detailed earlier (i.e., immunity to error through misidentification and immediate relevance for action)?"; and "How can we explain the acquisition of the first-person concept?"

So it seems that we must do more work to provide an account of the capacity to entertain "I"-thoughts. As we have seen, to simply posit a primitive, prereflective kind of self-consciousness that provides the basis for higher-order forms of self-consciousness but cannot be analyzed further does not prove satisfactory. Rather, we should attempt to provide such an analysis.

More recently, philosophers have made such attempts; in particular, some have argued that we can give an account of the essential features of full-blown self-consciousness in terms of simpler, more primitive, and notably nonlinguistic and nonconceptual forms of self-representation. In other words, it has been suggested that the ability to entertain "I"-thoughts or *de se* thoughts is independent of the mastery of the first-person pronoun and of concept possession in general and, moreover, that we need to appeal to certain types of nonconceptual representation to explain the emergence and the specific characteristics of "I"-thoughts, namely, their nonaccidental self-reference, their implications for action, and their property of being immune to misidentification. That is to say that the thought-language principle has been rejected in favor of a turn "away from language" and toward so-called nonconceptual forms of self-awareness (e.g., Hurley, 1997; Bermúdez, 1998; O'Brien, 2007; Vosgerau, 2009).

So the question we have before us now is whether theories of nonconceptual self-awareness can in fact succeed in providing a noncircular account of the ability to think "I"-thoughts: in other words, whether theories of nonconceptual self-consciousness can provide an analysis of what has so far been left unexplained both by the notion of prereflective self-consciousness and by the linguistic analysis of "I"-thoughts.

# 2 Nonconceptual Content

According to Bermúdez (1998), we can break the circularity of the linguistic approach to self-consciousness if we can show that there are thoughts with nonconceptual first-person content. "These are thoughts that, although first-personal in the sense that their content is to be specified directly by means of the first-person pronoun and indirectly by means of the indirect reflexive pronoun 'he*,' can nonetheless be correctly ascribed to creatures who have not mastered the first-person concept (as evinced in mastery of the first-person pronoun)" (Bermúdez, 1998, p. 45). Bermúdez believes he can achieve this by showing that there are thoughts with nonconceptual content that fulfill the three central characteristics of self-conscious thought that we identified in the previous chapter, namely, (1) nonaccidental self-reference, (2) immediate action relevance, and (3) immunity to error through misidentification. So these are thoughts that, although nonconceptual, nonetheless represent the self in the specific first-personal way that is distinctive of self-consciousness. He argues that we find examples of these in (visual) perception and somatic proprioception.

While I agree with Bermúdez that we should attempt to explain how self-consciousness can arise out of nonconceptual forms of representation—both to understand how self-consciousness can develop and because doing so will give us deeper insight into the structure of self-consciousness—I disagree that the way to go about this is by showing that the self is represented in the (nonconceptual) *content* of experience. This is because an approach that seeks to locate the origins of self-consciousness by arguing that the nonconceptual content of experience *represents* the self remains implicitly committed to the misguided subject–object model of self-consciousness and neglects the crucial insight expressed in Hume's "elusiveness" thesis. Moreover, it cannot explain the immunity to error through misidentification of paradigmatic forms of first-person thought. I propose instead that rather than trying to ground the ability to think "I"-thoughts in the representational *content* of experience, we should seek to ground it in the *mode* of experience. Moreover, I take it that we can break the circle associated with the linguistic approach without thereby claiming that there are nonconceptual forms of self-consciousness. While it is important to show how self-consciousness can arise out of nonconceptual forms of representation, this does not mean that these nonconceptual forms of representation must themselves be forms of self-consciousness. Rather, we can see them as necessary precursors to genuine, conceptual self-conscious thought. The central idea that I defend in the following is that while states with nonconceptual content can and often do contain *implicitly self-related information*, this should not be confused with *explicit self-representation*, where the latter is required for genuine self-conscious thought. Thus, to provide an account that grounds genuine, conceptual self-conscious thought in more primitive ways of representing the world, we must also tell a story of how an organism can get from implicitly self-related information to explicit self-representation.

Before we turn to this argument, it is first necessary to get a better understanding of the notion of nonconceptual content. Hence the task of this chapter is to introduce the notion of nonconceptual content and to motivate it in a way that is independent of the question as to whether it can successfully be applied to the problem of self-consciousness. Naturally, it is beyond the scope of this chapter to provide a comprehensive account of the notion of nonconceptual content and to deal with all the challenges that such an account would face. Rather, the aim here is to demonstrate that good reasons exist to posit nonconceptual forms of representation in general, and to sketch a positive characterization of nonconceptual content. Armed with this background, we can then turn to the question of whether and how the notion of nonconceptual content can be put to work to give an account of the ability to think "I"-thoughts, which I address in chapters 3 and 4.

The chapter begins with a brief characterization of nonconceptual content in opposition to conceptual content, and with an exposition of the basic motivation underlying the distinction between conceptual and nonconceptual content (sec. 2.1). This is followed by a discussion of the arguments in favor of nonconceptual content that are of particular relevance to our discussion (sec. 2.2). I then argue that we should understand the distinction between conceptual and nonconceptual content in terms of a difference in contents, rather than states (sec. 2.3). Finally, I provide the outlines of a positive characterization of nonconceptual content (sec. 2.4). In particular, I suggest that nonconceptual content is best understood in terms of procedural knowledge or "knowledge-how" rather than in terms of "knowledge-that." As I argue in chapters 5 and 6, this will have important implications for an explanation of how self-conscious thought emerges out of nonconceptual forms of representation. In particular, it suggests that we should distinguish between different levels of explicitness, and hence between different levels of self-consciousness.

## 2.1  Introducing the Notion of Nonconceptual Content

Before we start to think about the difference between conceptual and non-conceptual content, a few remarks about the notion of representational content are in order. The term "content," as many philosophers use it, and as I use it throughout the book, refers to the way in which some aspect or aspects of the world are presented to or grasped by the subject in her experience or thinking. The content of a mental state needs to be distinguished both from its vehicle (i.e., the mental state itself) and from its intentional object or objects (i.e., the aspect or aspects of the world that are being represented) (cf. Hopp, 2011). The way in which a subject grasps an aspect of the world can then be used to explain the subject's behavior. As we saw in the introduction, according to the principle of parsimony, we should only ascribe representational content to a creature when no explanation of the creature's behavior in terms of simple stimulus–response mechanisms is available. Moreover, we also saw in the introduction that representational content can (often, though, as we will see, not always) be evaluated in terms of correctness or accuracy. In sum, we can say that "representational states of mine have content in virtue of which they make the world accessible to me, guide my action, and (usually) are presented to me as something which is either correct or incorrect" (Cussins, 2003, p. 133).

Gareth Evans (1982) in his *Varieties of Reference* first introduced the notion of nonconceptual content into the analytical philosophy of perception.[1] In contrast to conceptual content, nonconceptual content is not composed of concepts. It is important to note that this is not a claim about the subpersonal level. Although the notion of nonconceptual content also plays an important role in the cognitive science literature that is concerned primarily with subpersonal information processing (Bermúdez, 1995), the notion as it is being employed here is intended to capture, as Cussins puts it, "how the person's perceptual experience presents the world as being (i.e., a genuine notion of personal level content). The notion of nonconceptual content is a notion which must ultimately be explained in terms of what is available in *experience*" (Cussins, 2003, p. 145; italics in original).

To assess the need for a notion of nonconceptual content, it is helpful to begin by briefly considering the role of concepts. What makes concepts philosophically interesting is their ability to account for the astounding systematicity, productivity, and rationality of human thought—in contrast to much of animal thought. Human thought is systematic, productive, and rational because it is composed of concepts that can be systematically decomposed and recombined and license rational inferences between thoughts. The same is not true for just any representation that suggests sensitivity to differences in the environment. Accordingly, although philosophy has no generally agreed-on theory of concepts, philosophers generally accept that conceptual content, at the least, consists of several components that can be systematically decomposed and recombined. In other words, conceptual content is generally considered to meet Evans's (1982) Generality Constraint.[2]

Again, the basic idea behind this constraint is that the possession of concepts enables a thinker to generate a potentially indefinite number of thoughts via the variable combination of concepts that the thinker possesses. More precisely, the Generality Constraint holds that conceptual mastery of the thought "a is F" implies the ability to think "a is G" for any property G of which the subject of the thought has a conception, and similarly to think "b is F" for any object *b* of which the thinker has a conception (Evans, 1982, sec. 4.3). For instance, a thinker who possesses the concepts RED and BALL and can entertain the thought "The ball is red," and also possesses the concepts GREEN and APPLE and can entertain the thought "The apple is green," should in principle be able to entertain the thought "The ball is green" and likewise the thought "The apple is red." Conceptual content is also thought to enable a subject to extend her thoughts on a matter beyond the current context. In other words, conceptual content is stimulus independent (Camp, 2009), or detachable from the present context (Hopp, 2011). Take again the concept of a ball. A thinker's possession of the concept will enable her not only to visually recognize balls in a variety of contexts but also to identify balls based on touch, to think about balls in their absence, or, in the case of a linguistic being, to recognize the word "ball" as an expression of the concept BALL when written or spoken.[3]

## 2.2 Why Do We Need Nonconceptual Content?

The main point of the previous section was that there is widespread agreement that conceptual content conforms to the Generality Constraint. In contrast, nonconceptual content is not subject to the Generality Constraint (e.g., Hanna, 2008; Heck, 2007; Meeks, 2006; Toribio, 2008). This is not to say that it cannot have any structure at all, but its structure is such that it does not contain components—such as objects, properties, and predication relations—that can be systematically decomposed and recombined in new contexts. So nonconceptual content is representational but does not consist of components that can be systematically decomposed and recombined, it is situation specific, and it is often restricted to a particular cognitive domain. As Hopp (2011) puts it, it is "non-detachable." Obviously much more needs to be said about how exactly nonconceptual content is to be characterized, and I discuss this subject in more detail in section 2.4. First, however, it is worth considering the question of why one might think that the notion of nonconceptual content is needed at all. Why not think instead—as conceptualists like McDowell (1994) and Brewer (1999) do—that all perceptual content is necessarily conceptual?[4]

As mentioned earlier, in specifying mental content, we are specifying the way in which aspects of the world are presented to a subject, or the way in which a subject takes the world to be, which, in turn, can then be used to explain a subject's actions. In other words, in specifying mental content, we are trying to capture its cognitive significance, or the role it plays in the organism's perception, thought, and action. And one important intuition that motivates many proponents of theories of nonconceptual content is that a subject's ways of apprehending the world are not always or not entirely constrained by the concepts a subject possesses. Why should this be so?

Various different arguments in favor of the notion of nonconceptual content have been brought forward and discussed in the literature. Some refer to the fact that our perceptual experience is extremely rich, presenting us with a huge number of objects and shapes that seem to outstrip our conceptual abilities (Chuard, 2007); others refer to the fact that the discriminable properties we perceive are much more fine grained than what our conceptual apparatus seems to allow for (Heck, 2000; Peacocke, 1992; Tye, 2006).[5] Against these arguments, it has been suggested that the richness and fineness of grain of these experiences can be represented conceptually, namely, by means of demonstrative concepts (McDowell, 1994). However, the nonconceptualist does not have to deny that perceptual experience is conceptualizable (e.g., by means of demonstrative concepts); rather, the claim is that the perceptual concepts we possess—including demonstrative concepts—are a function of the perceptual discriminations we are capable of making, and not the other way round (Bermúdez, 2007).

This point can be generalized to the claim that to be able to give a noncircular account of concept possession of a perceptual concept, we have to rely on the notion of nonconceptual content. According to Peacocke (1992), a concept is individuated by its application conditions. In other words, to explain what it is to possess a concept, we have to be able to provide the application conditions of the concept in question, that is, the circumstances in which it is appropriate to apply the concept. And to do so in a noncircular way, we cannot refer to the concept whose application condition is being elucidated (Coliva, 2003; Bermúdez, 2007). That is, in explaining the circumstances under which it is appropriate to apply a given perceptual concept, we must spell out how the perceptual content is given to the subject without referring to the concept in question to avoid explanatory circularity. So we need to give an account of how the perceptual content is given in terms of nonconceptual content.

Relatedly, Roskies (2008) has argued that it is hard to see how concept acquisition is possible without presupposing that we are capable of making the relevant perceptual discriminations *before* possessing the corresponding concepts. According to Roskies, the very fact that perceptual concepts have to be, and hence can be, learned suggests that before their acquisition, the content of the experiences we come to describe with the help of these concepts is, as it were, nonconceptual.

In short, to avoid explanatory circularity, we must give an account of what it is to possess a concept—in terms of spelling out its application conditions—that does not itself refer to the concept in question. Moreover, to explain how concepts can be acquired, we must assume that the world is presented to us in experience before the acquisition of the concepts we later come to employ to describe the experience. And this, in turn, suggests that we need to appeal to the notion of nonconceptual content.[6]

Arguably, this nonconceptual experiential content must also equip the subject with a sense of "primitive normativity" (Ginsborg, 2011a, 2011b). This is because concept possession is inherently normative (Peacocke, 1992; Ginsborg, 2011b). The possession or grasp of a concept implies knowledge with regard to the way the concept ought to be applied, that is, knowledge of the circumstances in which it is appropriate to apply the concept, as well as knowledge of the inferential transitions that are licensed by the concept. In other words, grasping a concept implies recognizing the rule that governs the application of the concept. For example, possession of the concept RED implies knowledge that the concept ought to be applied to any and only to red objects. Now, as Ginsborg (2011b) argues, if we want to explain how it is possible to acquire this kind of normative knowledge, we have to assume that there is a kind of "primitive normativity" or sense of appropriateness that enables a child (or an adult, for that matter) to get a sense of how things "belong together" during the process of learning to discriminatively respond to them. That is, to explain the possibility of acquiring the rule governing the application of a concept, such as RED, in a noncircular fashion, we cannot rely on the assumption that, say, in sorting red objects, the child was already following the rule in question. Rather, we need to assume that the child can acquire this rule through her activity of sorting red objects by coming to recognize in a primitive—that is, nonconceptual— way the appropriateness of what she is doing given her current context. Put differently, according to Ginsborg (2011b), we need to assume that the child has access to a primitive kind of normativity, which, in turn, makes possible the acquisition of the rules that govern the application of concepts (and thus makes possible the acquisition of conceptual knowledge).

The case in favor of nonconceptual content is further strengthened if we consider the possibility of intentional agency in creatures that do not possess conceptual abilities. As discussed earlier, it is commonly held that the possession of representations with conceptual content enables the subject to generalize, decompose, and recombine, and to apply the concept independently of a specific context. Arguably, many creatures, such as most animals and infants, cannot be attributed with these abilities. Nevertheless instances exist in which we have to attribute representational content to these creatures to explain their behavior—namely, instances in which we cannot appeal to a simple stimulus–response interpretation of their behavior, that is to say, in cases of intentional behavior. Intentional behaviors are those that need to be explained in terms of an agent's goals or desires in combination with the agent's representation of the environment. Psychological explanations like these work because they make intelligible why a given action is rational or "makes sense" from the perspective of the agent (Hurley, 2006). Such explanations ascribe to the agent in question precisely a kind of "primitive normativity" (Ginsborg, 2011a, 2011b).[7] That is to say, they ascribe to the agent in question the ability to respond to their environment in a way that makes this response appropriate to the agent's situation. A typical intentional explanation of my action of going to the kitchen and fetching some ice cream, for example, would be that I felt a desire for ice cream, I believed there to be ice cream in the freezer, and therefore I went to the kitchen to get it. Similarly, an intentional explanation of a dog digging a hole in the garden would be that the dog desires its bone, remembers it to be in the ground in the garden, and takes digging to be an adequate way to reach the bone and fulfill its desire. Insofar as a creature that does not possess concepts can nonetheless display intentional behavior (i.e., can display primitive forms of rationality), the observation of this behavior can justify the attribution of nonconceptual representational content to the creature. In other words, intentional agency occupies a middle ground between mere stimulus–response behavior and context-independent conceptual abilities (Hurley, 2006), and thus between the full-fledged normativity of conceptual thought and the lack of normativity of simple stimulus–response behaviors in the absence of a sense of appropriateness (Ginsborg, 2011a, 2011b).

An intentional agent can distinguish means from ends and can recognize that there may be different means to the same end, or the same behavior can be a means to different ends (Hurley, 2006). So intentional agents can—at least to some limited degree—flexibly adjust their behavior in different circumstances to bring about a desired goal; they can adjust their behavior in ways that render it appropriate to their current situation. In other words, an intentional agent is capable of instrumental reasoning, or practical rationality. Nonetheless intentional behavior can be restricted to a particular cognitive domain and need not conform to Evans's Generality

Constraint in the way that conceptual abilities do. As Hurley (2006) points out, behavioral flexibility can vary in nature and degree; it may be restricted to a specific domain, context, or function, for instance. Because of their domain specificity or restriction to certain contexts or functions, behaviors that display such a limited degree of flexibility need to be explained in terms of nonconceptual rather than conceptual representations.[8]

Consider the ability for instrumental reasoning displayed by some non-human animals. For example, Hurley (2006) discusses the following experiments involving the chimpanzee Sheba (Boysen & Berntson, 1995):

> Sheba was allowed to indicate either of two dishes of candies, one containing more than the other. The rule was: the candies in whichever dish Sheba indicated went to another chimp, and Sheba got the candies in the other dish. Sheba persisted in indicating the dish containing more candies at a rate well above chance, even though this resulted in getting her fewer candies. Boysen next substituted numerals in the dishes for actual candies. She had previously taught Sheba to recognize and use numerals "1" through "4." Without further training, Sheba immediately invoked the optimal selection rule, that is, she began to choose the smaller numeral at a rate well above chance, thereby acquiring the correspondingly larger number of candies for herself. The substitution of numerals seemed to make instrumental reasons for action available to her, as they seemed not to be when she was faced directly by the candies. When the numerals were again replaced by candies, Sheba reverted to choosing the larger number. (Hurley, 2006, p. 157)

Hurley suggests that this indicates that Sheba is able to apply instrumental reasoning, but her ability is bound to the symbolic context provided by the numerals. In this interpretation, Sheba is able to display intentional behavior in one context but is unable to generalize the underlying reasoning to a different—though relevantly similar—context. Sheba's behavior therefore seems to be an example of intentional agency, and thus of representational content, in the absence of conceptual skills.[9]

Further examples come from social contexts. For instance, Tomasello (1999) argues that nonhuman primates have a special ability to understand the sometimes highly complex third-party social relations between conspecifics. However, they have great difficulty understanding relations between objects. So their skills with regard to understanding relations do not generalize from the social to the nonsocial domain (Hurley, 2006). Similarly, chimpanzees seem to be able to apply basic mindreading abilities when confronted with a competitive context, but not in cooperative contexts (Call & Tomasello, 2008; see chap. 7 for further discussion). Also, as we will see in more detail in chapter 6, young infants (and some nonhuman animals) may display an implicit understanding of others as agents who act on objects and as partners sharing their mental states, but fail to combine their representations of others as actors *and* social partners (Spelke, 2009).

What these examples show is that rational intentional (i.e., primitively normative) agency can exist in the absence of conceptual abilities, which would allow agents to generalize their reasoning beyond the present context. Thus cases of context-bound and domain-specific reasoning like these are best accounted for in terms of nonconceptual representational content.

Finally, the abilities of infants to parse their visual environment into persisting objects have also been presented as evidence for the existence of nonconceptual representational content (Bermúdez & Cahen, 2008). For instance, some researchers have argued that infants are able to represent the world as being segmented into objects that exhibit certain regularities by detecting surface arrangements and surface motions (Spelke, 1990). In particular, Spelke (1990) argues that infants are able to form a representation of three-dimensional visual surface arrangements and motions with the help of processes that follow the basic principles of *cohesion, boundedness, rigidity*, and *no action at a distance*, thus reflecting basic constraints on the motion of physical bodies.[10] This can be tested by measuring infants' looking behavior toward different displays, where longer looking times are typically taken to indicate surprise. On the basis of such findings, some developmental psychologists conclude that infants possess the core *concept* of an object (e.g., Spelke, 1990; Carey, 2009), which they can employ in a primitive form of physical reasoning, and is gradually enriched as children gain additional physical knowledge about the world. Similar claims have also been made about what are taken to be core concepts in other domains, such as those involved in the cognition of spatial and causal relations, agency (involving, e.g., concepts such as animate-motion and inanimate-motion), and number cognition (e.g., Mandler, 2004; Carey, 2009). However, many theorists also think that the cognitive abilities in question are modular and domain specific. In contrast, as we have seen, it is widely accepted that conceptual representations must conform to the Generality Constraint, which means that conceptual thought is essentially domain general, systematic, and productive, enabling subjects to generate an infinite number of thoughts involving the concepts they possess. So insofar as the representations that are involved in many of the cognitive abilities found in infants are domain specific, these representations violate the Generality Constraint and thus cannot be considered forms of conceptual representation (Bermúdez & Cahen, 2008).

I discuss additional examples of such context-dependent and domain-specific representational abilities in chapters 5 and 6, where we will also consider how to make intelligible the transitions between implicit, domain-specific representations and explicit, domain-general representations. As I will argue in the following chapters, one reason for the domain specificity and context dependence of certain types of representation is that they are forms of *implicit* or *procedural* representation. For the information that is implicit in certain behavioral procedures to become accessible to other parts of the cognitive system, it needs to be recoded into a more explicit format (Karmiloff-Smith, 1996). This recoding leads to representations that are more abstract (and thus less rich and fine grained) but can be used for more general purposes. Because this recoding is likely to occur in degrees, there are going to be various levels of representational flexibility and generality (see chap. 5).

## 2.3 State Nonconceptualism or Concept Nonconceptualism?

The previous section demonstrated that a number of compelling arguments seem to favor the notion of nonconceptual content, most notably the argument that concept acquisition (at least in the case of perceptual concepts) is a function of perceptual discriminatory abilities and not vice versa, and the argument that we need to appeal to nonconceptual ways of representing the world if we are to make sense of the intentional behavior of creatures that cannot be credited with concept possession.

However, a number of authors (e.g., Byrne, 2003; Crowther, 2006; Heck, 2000) have argued that the claim that there are nonconceptual ways of representing the world admits of two different interpretations, named the *content view* and the *state view*, respectively. Roughly, according to the content view, nonconceptualism is a thesis about the *content* of experience. It holds that nonconceptual content is different in kind from conceptual content (in the ways specified in sec. 2.1). In contrast, according to the state view, nonconceptualism is a thesis not at all about different kinds of content but rather about different kinds of *states*. For the state nonconceptualist, it is consistent to hold that all content—whether perceptual or belief content—is conceptual, but nonetheless, for a subject to undergo a perceptual experience, she does not need to possess the concepts required to specify the content of her experience (Toribio, 2008).[11] One might think that the way in which the discussion has been presented in the previous section conflates this distinction.

However, the state view neither seems to be very plausible (Bermúdez, 2007) nor coherent (Heck, 2000; Toribio, 2008). Recall that the motivation for introducing a different notion of content in the first place was that it helps us to understand the ways a subject can grasp the world as being, and thus helps us to explain the subject's intentional behavior. The basic idea is that not all these ways require concepts, either because the subject in question, such as an animal or a prelinguistic infant, does not possess them, or because the representational state in question does not require them, as is arguably the case with perceptual experience. And it seems justified to appeal to the notion of nonconceptual content because subjects sometimes display intentional behavior, the explanation of which requires the ascription of representational content, although the subjects do not possess or entertain concepts. But, as Toribio (2008) has argued, the state view is of no use here. This is because the state view entails a notion of content that is unable to capture the way a subject grasps the world as being, for it characterizes content in terms of concepts although these concepts cannot be employed by the subject, and hence cannot help us explain the subject's intentional behavior. Accordingly, this undermines the justification for introducing the notion in the first place. Or it entails that a subject could exercise cognitive abilities (i.e., conceptual abilities) she does not possess, which is incoherent.

Moreover, as Bermúdez (2007) has argued, the state view is unable to account for the distinction between states that are concept dependent and those that are not. According to Bermúdez, it is simply hard to see what the basis for this distinction is, unless one refers to differences at the level of content. It seems natural to say, for example, that judgments are concept dependent while perceptual states are not, precisely because they involve different types of content. But this explanation is not available to the state nonconceptualist, for he would claim that both involve the same type of content. Referring to a difference in functional roles between judgments and perceptual states does not help either, for this just shifts the problem to a different level. After all, we would then need to explain why some functional roles are concept dependent while others are not, and it is equally hard to see how to come by such an account unless one refers to the different types of content involved.

In sum, it seems that the state view conflates the fact that we, as theorists, need to employ concepts to specify the content of a representational state with the content as it is given to the subject. Therefore I will assume that the only coherent interpretation of nonconceptualism is the content view. In other words, definitions of nonconceptual content that seem to be ambiguous between the state and the content view should always be interpreted in the sense of the content view if they are to make sense. Likewise, wherever I refer to nonconceptualism, the reader should assume that I refer to the content view, unless specified otherwise.

## 2.4 Nonconceptual Content as a Form of "Knowledge-How"

So far I have argued that we have good reasons to posit the notion of nonconceptual content, and we should understand the debate about nonconceptual content in terms of the content view, rather than the state view. However, as yet, we have only a negative definition of nonconceptual content, in opposition to conceptual content. Thus we should ask ourselves at this point whether it might not be possible to provide a more positive account of nonconceptual content. What might nonconceptual content be? How can it be further characterized?

I have argued that we should attend to the cognitive significance a given representational content has for an organism if we are to pay due respect to the fact that the notion of representational content serves a crucial explanatory role when it comes to understanding the intentional behavior of that organism. Our aim is to explain how experience presents the world as being to an organism, such that we can understand the organism's interactions with the world. And I claimed that to do so, it is sometimes necessary to appeal to the notion of nonconceptual content. So to provide a positive characterization of this content, we should consider the organism's possibilities for interacting with the environment. Accordingly, Cussins (2003) has argued that nonconceptual content presents the world not in terms of truth conditions—as conceptual content would do—but in terms of the affordances provided by the environment. In this view, to experience a sound as coming from behind, for example, is to take oneself to be in a particular position to locate its origin; and to perceive a cup as being at a particular distance is to take oneself to be able (or unable, as the case may be) to grasp it. Similarly, we might explain the differences between infant and adult perception by reference to the infants' abilities to interact with their environment, such as their ability to track the movement of objects, which change over the course of development, thus leading to changes in their perceptual content. This can also be expressed by saying that nonconceptual representations should be understood in terms of knowledge-how, or procedural knowledge, rather than knowledge-that.[12]

The distinction between knowledge-how and knowledge-that can be traced back to Ryle (1949, 1945), according to whom knowledge-how is an ability, which is in turn a set of dispositions. In contrast, in his view, knowledge-that is a relation between a thinker and a proposition.[13] Consider Cussins's example of riding your motorbike through London (Cussins, 2003, pp. 149–152). Given that you are a skilled rider, you will be able to wiggle with ease through the heavy traffic, avoiding road dividers and other obstacles, constantly adjusting your speed and balance in response to changing conditions. There is a distinct sense in which one could say that you know the speed at which you are traveling. But your speed is not presented to your experience in the form of a propositionally structured thought, such that you could express your speed in miles per hour. Rather, your speed is presented to you "as an element in a skilled interaction with the world" (p. 150), as a sense of how your environment would afford certain movements but not others. You possess an experiential, activity-based knowledge of your speed, a form of knowledge-how, which can come apart from the propositional knowledge-that of speed in terms of miles per hour. So you may possess knowledge-how without possessing knowledge-that, and vice versa. In fact, as Cussins points out, and as is commonly observed, a sort of tension even seems to hold between these kinds of knowledge. For instance, an unskilled rider may have to frequently check the speedometer and may have to infer the significance of the information thus gained in terms of her motorcycling activity, which will interrupt the flow of her riding. In contrast, a skilled rider may assess her speed and make the necessary adjustments without having to rely on this kind of inference. Likewise, a skilled football player does not have to pay attention to her every move; rather, the necessary movements come effortlessly and automatically, in contrast to those of a football novice. In fact, were such a skilled player to consciously attend to her moves so as to adjust them according to certain rules that are available to her in the form of propositions, her play would be interrupted, and she would most certainly make more mistakes and play less skillfully. At the same time, an avid reader of football instruction books may well have all the conceptual knowledge that there is to be had about playing football, she may have memorized all the rules of the game and all the different movements and actions required to become a skilled player, but she will nevertheless be unable to play, for she lacks the procedural, experiential knowledge required to be a skilled player. The advantage of having acquired this kind of experiential knowledge is that its links to action are direct, without any need for inference, thereby allowing for faster, more elegant actions. The disadvantage is that it is situation and domain specific and cannot provide the kind of generality and flexibility characteristic of conceptual representations. (I discuss this point in more detail in chap. 5.)

Similarly, many animals possess complex bodily skills that enable them to successfully engage in intentional interactions with the environment. In addition to complex bodily skills, other abilities also qualify as forms of knowing-how (see Bermúdez, 2003, p. 36). One example is empathetic reasoning, that is, the sort of reasoning about how people will behave that does not imply the application of an explicit theory to their behavior but rather consists in a basic form of empathy, or vicarious perception. I discuss this in more detail in chapter 6. Another example is trial-and-error reasoning, which is usually driven by a goal but does not imply the formation of explicit hypotheses about how the goal is to be achieved. We will encounter cases of such trial-and-error reasoning in chapter 5. Apart from the fact that these types of reasoning cannot easily be translated into a conceptual, propositional form, they share the property of being domain specific and situation dependent, because they are generally tied to the possibilities for action afforded by the particular environment.

So, according to the view proposed here, which goes back to Cussins but in similar form is also supported by other recent accounts (e.g., Hurley, 1998; O'Regan & Noë, 2002; Pettit, 2003; Ward, Roberts, & Clark, 2011; Hopp, 2011), a conceptual connection exists between the content of experience and an organism's abilities for intentional action—where these abilities include so-called "epistemic actions," such as grouping, sorting, and tracking objects or states of affairs. The basic point common to all these proposals is that the content of a given perceptual experience and its cognitive significance can only be fully captured by referring to the organism's abilities. This means that nonconceptual content can be understood as a form of procedural representation or knowledge-how. For instance, Hopp (2011) has recently argued—with reference to Husserl's notion of a "horizon"—that the nonconceptual content of perception is at least partly determined by the subject's "empty intentions" toward presently unseen parts or aspects of an object. These empty intentions can be interpreted as a form of sensorimotor knowledge about the ways in which presently unseen parts and aspects of an object would appear were the subject to change her position in relation to the object (see O'Regan & Noë, 2002). Similarly, Pettit (2003, p. 203) suggests that an object looks red insofar as "it manifestly enables you to sift and sort and track it in the red-appropriate manner," and the "ball that someone throws looks to be going fast so far as it manifestly elicits reaching *there* if I am to catch it, or ducking *now* if I am to avoid it." Likewise, according to Ward et al. (2011), the implicit knowledge of how to pursue and accomplish one's goals and intentions with regard to certain objects in one's environment determines the content of one's perceptual experience of that environment. For instance, my experience of the mug in front of me can be captured by referring to my abilities to grasp the mug and track it through space and time, and to my sensitivities toward changes in the mug's features. Likewise, my perceptual sensitivity toward a chair can be accounted for by my (implicitly) grasping that the chair affords seating. Note that this does not require that I must also be able to grasp that other objects equally afford seating. My experience can be entirely context bound and domain specific. Because of this, the possible actions that an agent implicitly takes herself to be able to perform can come apart from her explicit judgments, which can abstract away from the given perceptual situation. For instance, in the Müller-Lyer illusion (fig. 2.1), a subject will implicitly take herself to be able to perform actions, such as sorting, comparing, or touching, that imply a different length of the two lines, although (if the subject is familiar with the illusion) she will explicitly judge them to be of the same length.[14] In this view, therefore, a nonconceptual perceptual representation is accurate when the agent manages to engage appropriately with the world relative to her goals (i.e., when the world satisfies the agent's expectations regarding the possible actions she takes herself to be able to

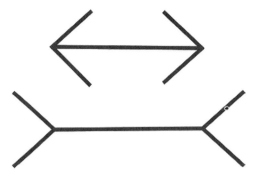

**Figure 2.1**
The Müller-Lyer illusion.

perform), while misrepresentations, such as illusions, occur when the world does not satisfy the agent's implicit expectations, and the agent does not manage to engage appropriately with the world (cf. Ward et al., 2011).

Note that this account implies that the "primitive normativity" associated with nonconceptual content is of a different kind than the normativity of conceptual thought. While conceptual thought is governed by the norm of truth—and thus is evaluated in terms of truth conditions—nonconceptual representation is governed by what Cussins (2003) calls "mundane normativity," which is the normativity of action guidance and skill.

However, you may wonder whether the account just sketched is appropriate for the explanation of complex intentional behaviors, such as those involved in more complicated practical inferences. Moreover, you might ask how nonconceptual representational content can "hook up" with concepts, such that conceptual abilities can be acquired on the basis of nonconceptual content, given that nonconceptual and conceptual content involve different kinds of normativity. These are important questions. Indeed, to answer them, we will need to move away from the simple dichotomy between nonconceptual and conceptual forms of representation toward a level-based approach that can make intelligible the transitions between relatively simple perceptual discriminations and basic forms of intentional behavior, toward more complex ways of engaging in practical reasoning and of representing states of affairs, and finally culminating in conceptual representations and theoretical reasoning. I provide the basis for developing such an account in chapter 5, where I discuss the difference between implicit and explicit representations and propose an argument for an account that assumes different levels of explicitness. For the moment, however, all I want to show is that it is necessary to posit forms of nonconceptual representation to account for the nature of perceptual experience and for the existence of intentional behavior in creatures that do not possess conceptual abilities, and to provide a rough outline of potential ways of specifying the content of these representations. According to the arguments presented here, nonconceptual representations should be understood as forms of "knowing-how" (or as forms of procedural knowledge), which is to say that to specify the content of these representations, one should take into account the abilities for action of the given organism.

## 2.5   Conclusion

The task of this chapter has been to motivate the notion of nonconceptual content independently of the question as to whether and how it can help us solve the problem of self-consciousness. Given that it seems plausible to assume the existence of nonconceptual content in general, we might be able to put this notion to use in the context of trying to solve the problem of self-consciousness that was exposed in the first chapter.

The discussion presented a variety of good reasons to assume the existence of nonconceptual content in general. The first argument showed that perceptual experience is richer and more fine grained than our concepts would seem to allow for. The second argument demonstrated that we need nonconceptual content to provide a noncircular account of the possession conditions of concepts and to explain the possibility of concept acquisition. And the third argument showed that nonconceptual content is needed to account for the fact that some beings, such as infants and animals, seem capable of displaying intentional behavior although we would not want to attribute them with conceptual abilities. Note that even if one does not agree that the content of perception is in principle richer and more fine grained than conceptual content, one can still believe that nonconceptual content is needed to account for concept acquisition and to account for the intentional behavior of nonconceptual creatures.

In advancing a positive characterization of nonconceptual content, the first point to note is that the difference between conceptual and nonconceptual content really is a difference at the level of content, not merely a difference in states (which may have the same kind of content). Further, the discussion showed that despite the many disagreements in this area, there is consensus that nonconceptual content—in contrast to conceptual content—is noncompositional and context bound, which is to say that it does not conform to the Generality Constraint. I also argued that nonconceptual content should be understood in terms of knowledge-how rather than knowledge-that. We will return to these points in chapter 5, where I argue that the domain specificity and context dependence of certain nonconceptual types of representation have to do with the fact that these representations are encoded in a procedural, implicit format, which is distinguished from a more abstract, explicit way of encoding information that can be applied more flexibly across domains. Implicit and explicit representations can come apart and may sometimes even conflict with one another. Chapter 5 will also demonstrate how procedural forms of representations can gradually be transformed into more explicit forms of representation. I argue that this transformation occurs via a process of representational redescription across several levels of explicitness, which explains why we find various degrees of flexibility when considering the intentional behavior of pre- and nonlinguistic beings.

# 3 Self-Representationalist Accounts of Nonconceptual Self-Consciousness

Recently, some philosophers have attempted to apply the notion of nonconceptual content to the problem of self-consciousness, and to use this notion to argue in favor of forms of self-consciousness that rely neither on self-reflection nor self-identification, nor on the ability to use the first-person pronoun (e.g., Bermúdez, 1998; Hurley, 1997; O'Brien, 2007). Having discussed reasons that speak in favor of the existence of nonconceptual content in general in the previous chapter, I turn now to theories of nonconceptual self-consciousness.

The chapter begins by contrasting self-representationalist and non-self-representationalist (or no-self) approaches (sec. 3.1). Self-representationalist approaches are my primary target in this chapter, and I discuss non-self-representationalist approaches in more detail in the next chapter. This chapter then discusses arguments in favor of the claim that the self is represented in the nonconceptual content of experience. I first consider the self as being represented in the content of perception (sec. 3.2) before considering bodily experience as a form of nonconceptual self-awareness (sec. 3.3). Section 3.4 shows that these arguments are misguided, and the self is not in fact part of the representational content of experience. Accordingly, perception and bodily experience are not forms of self-consciousness—though they are among the foundations of self-consciousness. Section 3.5 provides a conclusion, and section 3.6 discusses the relevance of recent neuroscientific studies dealing with self-consciousness.

## 3.1 Self-Representationalist versus Non-Self-Representationalist Accounts

There are two ways of going about giving an account of conceptual self-consciousness based on nonconceptual forms of self-awareness. As mentioned in the introduction, we can distinguish self-representationalist from non-self-representationalist accounts of nonconceptual self-awareness. Before turning to a discussion of these accounts, let us briefly recall the motivation for introducing the notion of nonconceptual self-consciousness.

Bermúdez (1998) argues that we need nonconceptual self-consciousness to solve what he has called the "paradox of self-consciousness." This paradox arises from two core assumptions: (1) to understand self-consciousness, it is necessary to understand a person's ability to have "I"-thoughts, where "I"-thoughts are thoughts that nonaccidentally refer to the thinker of the thought; and (2) to understand the capacity to think "I"-thoughts, it is necessary (and sufficient) to analyze their canonical linguistic expression, that is, the use of the first-person pronoun. But the ability to use the first-person pronoun presupposes the ability to think "I"-thoughts, for a speaker who is using the first-person pronoun correctly has to understand that it refers to herself, and this understanding is itself an "I"-thought. Hence we seem to have an explanatory circle. Moreover, it seems that we cannot explain how the ability to use the first-person pronoun can be acquired without presupposing the existence of ways of thinking about oneself that already rely on the possession of this ability. However, as Bermúdez points out, for any ability that is psychologically real, we should—at least in principle—be able to provide an account of how it can be acquired.

A possible solution to the paradox identified by Bermúdez consists in rejecting the second of the two assumptions mentioned in the previous paragraph. To do so, one can attempt to demonstrate that a primitive capacity to think "I"-thoughts does not require the capacity to refer to oneself linguistically—or conceptually, for that matter. In other words, one can attempt to demonstrate the existence of nonconceptual forms of self-representation that (1) are self-referring, (2) have immediate implications for action, and (3) are immune to error through misidentification. Alternatively, one can deny that the nonconceptual content of experience represents or refers to the self, and argue that it can nonetheless ground conceptual "I"-thoughts in a way that makes intelligible how "I"-thoughts can possess the features just listed and how the self-concept can be acquired.

Bermúdez's account falls into the former category. He argues that the self is part of the representational content of experience (e.g., of perception and proprioception), and that this fact explains why perception and proprioception fulfill the three criteria mentioned in the previous paragraph.

Thus, in his account, there are primitive, nonconceptual forms of genuine self-consciousness, namely, perception and somatic proprioception, which provide the basis for other, more complex forms of "I"-thought.

Other accounts, such as those proposed by Peacocke (1999), O'Brien (2007), or Recanati (2007), do not assume that the self is explicitly represented in the content of experience; rather, in these views, it is part of the *mode* of experience. Thus these authors view the content of experience as being "self-less." Nonetheless, in these accounts, experience can ground conceptual self-ascriptions, which in turn fulfill the three criteria mentioned earlier.

The task of this chapter is to examine the first type of account and to show that it has serious shortcomings. In particular, this account fails to demonstrate that there are forms of nonconceptual representation that fulfill all three criteria of self-consciousness, namely, (1) self-reference, (2) unmediated action relevance, and (3) immunity to error through misidentification. This is because it fails to distinguish between implicitly self-related information and explicit self-representation. As a result, this account mischaracterizes the content of experience and remains—against Hume's insight—committed to the subject–object model of self-consciousness.

As I argue in the next chapter, "no-self" accounts do not suffer from these problems. However, existing versions of them fall short in that they do not explain how it is possible to get from implicitly self-related information to explicit self-representation and self-reference.

### 3.2 The First Person in Perception

Proponents of the self-representationalist view take visual perception to be one of the most fundamental forms of nonconceptual self-consciousness. The basic idea is that although *prima facie* visual perception is predominantly directed outward, there is a sense in which the information received through perception is necessarily also about the perceiving organism itself. This has to be so for the information in question to become relevant for action and self-maintenance.

The argument that visual perception necessarily contains information about the perceiving subject goes back to Gibson's (1979) theory of ecological optics.[1] Based on Gibson's (1979) theory, Bermúdez (1998) argues that the structure of visual perception contains information about the self in various ways: On the one hand, there are self-specifying structural invariants such as the boundedness of the visual field and the occlusion of parts of the visual field by various parts of the body. This means that your range of vision is limited, and your visual field is bounded, which is to say that you always see things from your egocentric perspective. Moreover, various parts of your body, such as your limbs, your nose, and so on, occlude specific parts of your visual field. Thus the outline and contours of the subject's body are present in the field of vision as "self-specifying structural invariants" (Bermúdez, 1998, p. 109). This suggests that—contrary to the view that holds that vision only delivers information about the external world—the self is part of the content of visual perception, namely, in the form of self-specifying information.

On the other hand, there are differing patterns of flow in the optic array as one moves through the environment and the environment around one moves, and the relations between the variant and invariant features of this flow enable the perceiver to distinguish between her movement in the world and the movement of the world around her. This is called visual kinesthesis. An example of a visually kinesthetic invariant in the optical flow of a perceiver that is moving through the environment is the point that is being approached by the perceiver. This target point is the vanishing point of the optical flow, which remains stationary, so that the subject perceives the overall pattern of the optic flow as changing relative to this point.

Vision is kinesthetic in that it registers movements of the body just as much as does the muscle-joint-skin system and inner ear system. Vision picks up both movements of the whole body relative to the ground and movement of a member of the body relative to the whole. Visual kinesthesis goes along with muscular kinesthesis. The doctrine that vision is exteroceptive, that it obtains "external" information only, is simply false. Vision obtains information about both the environment and the self. (Gibson, 1979, p. 183; cited in Bermúdez, 1998)

According to Bermúdez, the fact that visual perception contains information specifying the movement of the subject is another demonstration of the point that the self does have a place in the content of perception.[2]

Finally, the perception of affordances—that is, properties of objects in the environment that relate to the abilities of the perceiver—provides the subject with information not only about the objects that are being perceived but also about the possibilities for action that these objects afford, and thus about the subject herself. The perception of affordances was already mentioned in the discussion of nonconceptual content in the previous chapter, where I argued that to determine the nonconceptual content of perception, we should refer to the subject's abilities of interacting with her environment. That is to say, the environment is always perceived in terms of the kinds of interaction it affords. According to Gibson, these affordances are perceived directly as higher-order invariants as a subject moves around the world. For example, I immediately perceive the coffee mug in front of me to be within my reaching distance, and I know that I can safely sit down in my office chair, as I immediately perceive it to be stable enough to afford this action. The perception of affordances is another way in which self-specifying information is obtained, for the idea of an affordance is that the perception of the environment provides the subject not just with information about the objects that are perceived, but also with information about the subject's own possibilities for interacting with the objects in question.

The crucial point in all of this is that the structural invariants of the visual field, visual kinesthesis, and the perception of affordances provide the organism not just with information about the environment but at the same time with information about the organism itself. In representing properties of the environment, the perceiver receives information about her position in the world relative to other objects, about whether she is moving or stationary, and about her possibilities for interaction with the environment. Moreover, this information is not just contingently but necessarily self-related, for the perceiver can only receive information about herself, not about another organism, in this way. In other words, the self-related information in question relies on ways of gaining information that are specific to the self. Thus no self-identification is required; rather, the information that the subject receives about herself in terms of affordances, movement, and her spatial relation to objects in the environment is such that it necessarily refers to the subject herself.

Proponents of theories of nonconceptual self-consciousness have taken this to be an argument for the fact that these kinds of representations are both self-referring and immune to error through misidentification (Bermúdez, 1998; see also Vosgerau, 2009). Moreover, ecological perception also has immediate implications for action: for instance, my interaction with the objects in my environment is directly related to what kind of affordances I perceive them as offering me (Bermúdez, 1998, p. 117). An example of this, discussed by Bermúdez (1998), is the reaching behavior of infants. Field (1976) found that the reaching behavior of fifteen-week-old infants is adjusted to take into account the distance between an object and the infants themselves. Infants reach less often for an object as the distance increases. According to Bermúdez, this is best explained by saying that the "infants perceive the affordance of reachability in a way that has immediate implications for their behavior" (Bermúdez, 1998, p. 127).

Finally, the content of visual perceptions that deliver this kind of information can be classified as nonconceptual. First, as we have just seen, this kind of content can be ascribed not just to mature humans but also to infants and animals that are capable of visual perception and interaction with the environment. Second, we saw in the previous chapter that, even in the case of mature humans, there are ways of specifying perceptual content that do not require us to refer to the conceptual abilities of the perceiving subject; in fact, we may not even be able to capture this content adequately in terms of concepts. That the content of perception necessarily contains affordances and other types of action-guiding information also confirms the analysis given in the previous chapter, according to which the nonconceptual content of perceptual experience should be understood in terms of knowledge-how, that is, in terms of the possibilities for action it makes available (see sec. 2.4).

So what perception, according to Bermúdez, provides us with is a form of nonconceptual *first-person* content. In other words, it is content that fulfills all three criteria of self-conscious thought, namely, (1) self-reference, (2) unmediated action relevance, and (3) immunity to error through misidentification. Thus in Bermúdez's view we here have an example of a basic form of nonconceptual self-consciousness.

The arguments brought forward by Bermúdez are similar to arguments presented by Hurley (1997) with regard to what she terms "perspectival self-consciousness." As demonstrated by the examples discussed earlier in this section, and by the considerations presented in chapter 2, there is a systematic interdependence between perception and action, and vice versa. For example, if I move my head, my visual field will change, so that I now see things that were not part of my visual field before my movement. Likewise, the distance and angle I perceive the cup in front of me to be from the position of my hand will determine how far I stretch my arm and in which way I position my fingers to grasp it. Because of this interdependence, Hurley argues, an agent needs to possess the ability to use information that is about herself to perform intentional actions. In this sense, according to Hurley, having a perspective—which is a necessary feature of perception, for all perception is perception from a perspective, based within an egocentric frame of reference—necessarily involves being self-conscious.

If Bermúdez and Hurley are correct in the part of their analysis that characterizes the content of visual perception as a form of nonconceptual self-consciousness, then we have a model that demonstrates the ability to think "I"-thoughts without presupposing the possession of the first-person concept or the ability to use the first-person pronoun. In other words, we have a model of nonconceptual first-person thought that can provide the basis for the ability for conceptual first-person thought. However, as I argue in the following, this conclusion is drawn too hastily, for serious doubts can be cast on the thesis that ecological perception amounts to explicit self-representation, and self-representation is required for self-consciousness. Indeed, the claim that the self is part of the *content* of ecological perception will turn out to be misguided. Before we turn to these arguments, I present another alleged form of nonconceptual self-consciousness, namely, bodily awareness.

### 3.3   The First Person in Bodily Awareness

In addition to ecological perception, proponents of theories of nonconceptual self-consciousness generally regard proprioception as a primitive and foundational form of self-consciousness. Proprioception provides the organism with an ongoing stream of information regarding the state of the body, such as its position in space and posture, its muscular activity and strength, and so on. It seems obvious that this information is about the proprioceiving subject, and that it is necessarily so.

One of the distinctive features of somatic proprioception is that it is subserved by information channels that do not yield information about anybody's bodily properties except my own (just as introspection does not yield information about anybody's psychological properties except my own). It follows from the simple fact that I somatically proprioceive particular bodily properties and introspect particular psychological properties that those bodily and psychological properties are my own. (Bermúdez, 1998, p. 149)

So proprioception informs the subject about *her* bodily properties, such as the position of her own limbs—it cannot provide information about the limbs of anyone else—and does so in a way that is of immediate relevance for action. Moreover, it is a form of nonconceptual representation, for an organism does not have to be in a position to entertain the concepts required for specifying the content of its proprioceptive states to receive and process proprioceptive information. Thus, according to Bermúdez, proprioception should be regarded as a form of nonconceptual self-consciousness.

Anderson and Perlis (2005) have recently proposed an account that is based on similar considerations. They are trying to naturalize what they call the "essential prehension"—in analogy to Perry's notion of the "essential indexical"—that is, the prereflective, identification-independent grasp on the self that can ground full-fledged self-consciousness. They argue that proprioception can play that role. Similar to Bermúdez, they hold that proprioception can account for immunity to error through misidentification, for it necessarily delivers information about the proprioceiving subject. Further, they hold that somatic proprioception provides the organism with a self-representation that is sufficient to guide and motivate action, thus accounting for the immediate implications for action, which, as we saw in chapter 1, is an essential feature of self-consciousness. Thus proponents of theories of nonconceptual self-consciousness argue that proprioception is a form of nonconceptual representation that delivers information that is (1) about the self, (2) action guiding, and (3) immune to error through misidentification. We will see in the following why this is only partly correct, that is, why theories of nonconceptual self-consciousness are problematic.

Notice, however, that although Bermúdez emphasizes the discussion of proprioception, bodily awareness is not determined by proprioception alone, but should rather be understood as the result of a multimodal integration mechanism. In addition to proprioception, our sense of our body is also determined by the information we receive from other modalities, such as touch and vision. Moreover, proprioception can only guide one's behavior if it is properly integrated with perceptual information about the environment—and likewise, perceptual information can only guide action if it is integrated with proprioceptive information.

Indeed, according to Anderson and Perlis, proprioception provides a spatial "self-context" that can be integrated with other self-specifying information, such as visual information, thereby enabling the organization of various kinds of information into a coherent schema, allowing for the control and guidance of action, and constraining the content of our bodily experience.

Because of its specialized structure, the somatoceptive system has a firm grasp on the self—what we call the essential prehension—in virtue of which it produces self-specifying representations with just the right content and connections to make an information bridge to, and allow the proper organization of, other information about the self. More importantly, "SR*" [the self-referential mental token], in virtue of its grounding in somatoception, has the right connections with our action-guiding systems to account for the special motivational properties of the information organized under it, and the apparent self-directedness of certain actions. (Anderson & Perlis, 2005, p. 324)

This means that somatic proprioception delivers the organism information about itself (i.e., self-referring information), which is integrated with information gained through, for instance, visual (ecological) perception or touch, thereby guiding self-directed, as well as world-directed, actions of the organism. Thus, rather than focusing on proprioception in isolation from other sensory modalities, we should focus on the multimodal representation of the body, which integrates proprioceptive information with visual and other types of sensory information about the body, thereby presenting the body as the "point of convergence of action and perception" (Legrand, 2006; on the multimodality of bodily experience, see also de Vignemont, 2007, 2012). Accordingly, in the following, I use the terms "bodily awareness" or "bodily experience," rather than the term "proprioception."

It seems, then, that the body schema provides us with the basis for a sense of our body by integrating information from different modalities. Moreover, the spatial bodily representation provided by the body schema is self-specific: it necessarily contains information about the subject's own body, posture, size, and strength of limbs; we cannot receive information about the body of another in this way.

This seems to imply that bodily experience is immune to error through misidentification. If I experience a bodily sensation "from the inside," I can be wrong about what kind of experience exactly it is, or where it is located in body-space—as illustrated by the rubber hand illusion[3]—but I cannot be wrong about whose body it is that I am experiencing (de Vignemont, 2012).

Thus, or so proponents of self-representationalist theories of nonconceptual self-consciousness argue, bodily experience is a form of nonconceptual self-consciousness. As we have just seen, bodily experience is necessarily experience of one's own body, and it is closely related to one's possibilities for action. Hence one might conclude that bodily experience fulfills all three of the criteria that were established for "I"-thoughts: (1) self-reference, (2) relevance for action, and (3) immunity to error through misidentification.

To sum up, I argued that ecological perception, as well as the information delivered through multimodal bodily experience, enables us to move and interact with the world without requiring the ability to entertain propositionally structured thoughts about the environment. Proponents of theories of nonconceptual self-consciousness argue that ecological perception and bodily experience therefore constitute primitive forms of self-consciousness because they (a) deliver the organism self-specifying information that is relevant for action and (b) are immune to error through misidentification because they necessarily deliver information about the organism itself. Ecological perception delivers information that is necessarily about one's own spatial position relative to other objects in the environment and about one's own possibilities for interaction with these objects; bodily experience delivers information that is necessarily about one's own body and does so in a way that presents the body as a system of possible movements. If I perceive the mug in front of me to be within reaching distance, no question can arise as to *who* it is that can reach the mug. Likewise, if I have a bodily experience of my legs being crossed, there is no question as to *whose* legs are crossed; perception and bodily experience do not require any self-identification. That seems to suggest that they are immune to error through misidentification. And it is obvious that this information has immediate implications for my actions, for instance, my grasping the mug. Moreover, perception and bodily experience are nonconceptual forms of representation, since they do not require concept possession and can clearly be attributed to creatures that cannot be attributed with concept possession, such as animals and infants. It seems, then, that we have successfully established the existence of nonconceptual forms of self-consciousness.

However, in the following, the conclusion that, by virtue of their essentially self-related nature, ecological perception and bodily experience are to be considered forms of self-consciousness—in a self-representationalist sense—will be called into question. Perception and bodily experience do not represent the self. Hence, or so I will argue, the analysis presented here does not succeed with regard to providing an account of the ability to have "I"-thoughts independently from conceptual and linguistic abilities. I will show that nonconceptual content does provide the organism with

implicitly self-related information, but this does not amount to self-representation or self-reference, which are required for "I"-thoughts. Moreover, I
argue that the nonconceptual content of perception and bodily experience
cannot be considered to be immune to error through misidentification.
So self-representationalist theories of nonconceptual self-consciousness are
unable to accommodate (1) and (3) of the criteria for self-consciousness
mentioned earlier.

### 3.4   Arguments against the Self-Representationalist Approach

There are two main reasons to think that self-representationalist theories
are misguided and that non-self-representationalist, or "no-self," theories
are better suited to solve the problem of self-consciousness. The first is that
self-representationalist theories mistakenly aim to establish that the self is
explicitly represented in experience. However, as I argue in the following,
the self should instead be understood as an "unarticulated constituent" of
the content of experience. Indeed, positing that the self is explicitly represented in perception and bodily awareness puts an unnecessary and implausible cognitive burden on organisms capable of basic interactions with the
environment. Moreover, the claim that the self is part of the explicit content of perception and bodily awareness misrepresents the phenomenology
of perceptual and bodily experience and reveals an implicit commitment to
the subject–object model of self-consciousness. Thus it cannot provide the
correct analysis of the notion of prereflective self-consciousness that we are
seeking (see chap. 1). The second, closely related reason is that self-representationalist theories cannot show that perception and bodily experience
are immune to error through misidentification, nor are they well suited to
explaining how allegedly self-representational states such as perception and
bodily awareness can provide the basis for judgments that are immune to
error through misidentification.

### 3.4.1   Implicitly Self-Related Information versus Explicit
### Self-Representation

This section presents an argument that shows that so-called nonconceptual
forms of self-consciousness do not represent the self, for they lack an explicitly self-referring component. Accordingly, they should not be considered
forms of genuine self-consciousness. The core of this argument is that while

ecological perception and bodily experience contain implicitly self-related information, this does not amount to self-representation.

As was shown in the previous sections, perception and bodily experience provide the organism with information that is essentially self-related. Examples of this information include information regarding the distance between the organism and an object in its visual field; the position of the organism's limbs; whether the organism is moving or not; and the organism's possibilities for interacting with the environment. However, as I will now show, this information is not explicitly represented *as* being about the organism; rather, the fact that this information concerns the self remains implicit in the experience. In other words, we can think of the self as being an "unarticulated constituent" of experience.

Here is the argument in a nutshell: As Perry (1986, 1998a) has convincingly argued, facts that are provided by the context do not figure as part of the explicit representational content of an utterance or of an intentional state (see also Recanati, 2007). Moreover, it would put an unnecessary cognitive burden on the organism to represent itself explicitly if such explicit representation is not required for successful interaction with the environment. However, self-consciousness requires explicit self-representation. Hence, while perception and bodily experience are instances of conscious experience, they do not constitute forms of *self*-consciousness.

To put it differently: all sentient beings are subjects of experience, and they experience the world from their own egocentric perspective. Because of this, perceptual content necessarily contains self-related information, which the organism must use to interact with the environment. However, not all subjects of conscious experience also have explicit self-representations or think of themselves as themselves. Let me illustrate this thought by means of an example: The squirrel that is sitting on a branch of the walnut tree in front of my window and is about to jump to a neighboring branch has a certain perspective and is acting from this perspective. Assuming that it is acting intentionally, that is, assuming that we can ascribe representational content to the squirrel, one possible explanation of the squirrel's behavior may be something like the following: The squirrel sees some walnuts on the neighboring branch, it wants the nuts, and consequently it jumps onto the branch. But to behave in this manner, the squirrel does not need to have any explicit representation of itself as an individual agent or of having a particular perspective; it only needs to represent the tree and the nuts. To be sure, the content of the squirrel's tree representation will be partly determined by the squirrel's abilities for interacting with the tree, by its distance from the neighboring branch, and its abilities to jump, for example. That is to say, the squirrel will perceive the tree in terms of the kind of actions it affords, which are obviously relative to itself. Nonetheless the squirrel does not need to represent its distance to the tree and its perspective on the tree *as such*. Rather, that there is self-specifying information contained in the squirrel's perception of the walnut tree can remain implicit; the content of perception does not have to be presented to the squirrel *as* being self-related. Implicitly self-related information is sufficient for action guidance; it need not be represented *as* being self-related to be used by the squirrel's action-guiding systems.

We can spell this out further with the help of distinctions that were originally introduced by John Perry. To frame the foregoing example in Perry's terminology, we might say that the squirrel has "agent-relative knowledge," that is, knowledge from the perspective of a particular agent; but it does not have a notion of itself *as* an agent and hence does not have what Perry calls "self-attached knowledge" (Perry, 1998b).[4] While it may seem to be the case that agent-relative knowledge involves explicit reference to the agent, this is not so. It may seem to us that agent-relative knowledge involves self-reference because to get to the nut, the squirrel has to locate the branch holding the nut relative to the squirrel's own position; it has to know where the branch is relative to its own body to successfully perform the jump. That is, it has to have a grasp of the agent-relative role that the branch plays in the squirrel's means-end reasoning of how to get to the nut. And this, in turn, seems to require a representation of the agent itself.[5] Similarly, as Perry observes, when using the computer's keyboard to type these sentences, "I have to move my fingers a certain distance and direction from *me*. It isn't enough to know where the buttons were relative to one another, or where the [keyboard] was in the building or the room. I had to know where these things were relative to *me*. It seems then, that these basic methods [i.e., methods of finding out about the agent-relative roles that objects in our environment play] already require me to have some notion of myself" (Perry, 1998b, p. 86). This intuition is what seems to be driving the idea that the types of nonconceptual content discussed in the previous sections amount to self-consciousness.

But this view is mistaken. Having a grasp of the agent-relative role of the objects around oneself is not the same as having a representation of oneself as an agent. That is, agent-relative knowledge is not sufficient for self-attached knowledge, or for self-consciousness in the sense of having the ability to entertain "I"-thoughts. We can see this when we consider whether, in expressing the content of the squirrel's experience or the content of our experience when typing on our keyboards, we ought to use the first-person pronoun. This is not the case. Indeed, if we were to attempt to capture the squirrel's nonconceptual experience (which would require using concepts on our part, of course) the expression "nut in front" would arguably give a better characterization than the expression "nut in front *of me.*" After all, nothing in the squirrel's representation explicitly informs it about the food being in front *of itself.* The squirrel's representation does not need to contain a self-referring component to be action guiding. Likewise, to capture my perception of the keyboard, I might use the expression "keyboard in front at such-and-such an angle" rather than "keyboard in front of *me.*" This is because "sometimes all of the facts we deal with involving a certain *n*-ary relation involve the same object occupying one of the argument roles. In that case, we don't need to worry about that argument role; we don't need to keep track of its occupant, because it never changes" (Perry, 1998b, p. 87). In the case of agent-relative representations, the argument role of the agent itself always remains the same. All objects that the squirrel perceives are perceived from its egocentric perspective, in relation to itself. Consequently there simply is no need for the squirrel to keep track of whose perspective is represented in its perception because it never needs to distinguish its own perspective from that of others, and so the question of whose perspective is represented never arises for the squirrel (cf. Becker-mann, 2003).[6]

So although the squirrel's perception of the nut *concerns* the subject of perception, since the nut is perceived relative to the squirrel's perspective, it is not *about* the subject. And the same is, of course, true of a subject's bodily experiences. These may be expressed, for instance, by "itch in right foot" rather than "itch in *my* right foot." Thus perceptual or bodily experiences are not about the self; they are not self-intentional in the way "I"-thoughts are, although they necessarily contain implicitly self-related information. Put briefly, the content of perceptual and bodily experiences does not refer to the self. What it refers to are the objects of experience, such as the tree branch in the squirrel's case, or the keyboard, or the itch in a foot, in the human case.

A related point is made by Campbell (1994) in his discussion of egocentric spatial representation. As Campbell puts it, "The egocentric frame used in vision employs monadic spatial notions, such as 'to the right,' 'to the left,' 'above,' 'in front,' and so on, rather than relational notions, such as 'to my right,' 'above me,' 'in front of me,' and so on" (Campbell, 1994, p. 119). So we should not think that the content of visual experience itself already employs the first person. Rather, it provides the organism with "monadic," but implicitly self-related, information that, once conceptualized, can provide the basis for first-personal thought.[7]

As mentioned earlier, we may also frame this, to use another term from Perry, by saying that in the cases just described, the subject itself remains an "unarticulated constituent" of the content of its perception.[8] According to the theory of unarticulated constituents, "we don't articulate the objects we are talking about, when it is obvious what they are from the context" (Perry, 1998a, p. 11). One example that Perry uses to illustrate this idea is the well-known case of the Z-landers (Perry, 1986; reprinted in Perry, 2000). Residents of Z-land never get any information about the weather anywhere else, and they are not aware of the existence of other places. When they say, "It is raining," this statement is obviously relevant to the weather in Z-land, and they are able to act accordingly (by, e.g., taking an umbrella when leaving the house), although Z-land does not figure explicitly in the content of the utterance. To determine the truth conditions of the sentence "It is raining," we need a location (in this case Z-land). Nonetheless Z-land does not figure as an explicit component in the utterance (for it is supplied by the context), and neither need it be explicitly represented for the statement to be understood. In contrast, when I think or assert, "It is raining," depending on the context, I might have to specify the location I am referring to. It might be the case, for example, that I was just talking about my sister, who happens to live in another country and just called to complain about the weather where she is. In this situation, I might have to disambiguate my assertion by saying, for instance, "It is raining *here*" or "It is raining *there*," since I might be referring either to our current location or to hers. In contrast to us, Z-landers do not need to consider the weather in other locations when thinking or speaking about weather; as a matter of fact, they are not even able to consider other locations. Thus their thoughts about weather

(necessarily) concern Z-land insofar as they lead to behavior that is appropriate to the weather conditions in Z-land (e.g., taking an umbrella upon
thinking, "It is raining"), but Z-land does not have to be represented for
this to hold. There simply is no need for an explicit component referring to
Z-land in their thoughts for the connection to Z-land to be secured.[9]

Similarly, a being that never learns to distinguish its own perspective
from that of others does not need to explicitly represent its own perspective
in its perception and thinking. In fact, even for beings like us, who do have
a sense of our own perspective, this perspective will generally be absent
from the explicit representational content of our perceptual experience, for
it would put an unnecessary burden on our cognitive system to always represent it. As Perry himself points out:

> What each of us gets from perception may be regarded as information concerning
> ourselves, to explain connections between perception and action. *There is no need for
> a self-referring component of our belief, no need for an idea or representation of ourselves.*
> ... The eyes that see and the torso or legs that move are parts of the same more or
> less integrated body. And this fact, *external to the belief,* supplies the needed coordi-
> nation. The belief need only have the burden of registering differences in my envi-
> ronment, and not the burden of identifying the person about whose relation to the
> environment perception gives information with the person whose action it guides.
> (Perry, 2000, pp. 182–183; italics mine)

Accordingly, perception and bodily experience can represent properties and
states of oneself without representing the subject of these states. Indeed, it
is precisely because the self-related information implicit in perception and
bodily experience necessarily concerns oneself that no representation of
the perceiving and proprioceiving subject is required to guide the subject's
actions. If so, it would be unnecessarily cognitively demanding to assume
that the self is explicitly represented in experience whenever an organism
is interacting with the environment. Following the principle of parsimony
with respect to cognitive explanations, we should instead favor the less-
demanding explanation, according to which, while there is implicitly self-
related information, there is not explicit self-representation in the content
of perception and bodily experience. In other words, contrary to what self-
representationalist theories of nonconceptual self-consciousness seem to
suggest, no self-referring component in the experiential state is required to
secure the connection between perception and action. What these theories
seem to suggest is that since perception and bodily experience are relevant

for guiding the intentional actions of an organism, they must be about the self; in other words, they must be self-referential. But as we have just seen, this is not so. Perception and bodily experience can guide an organism's intentional behavior in virtue of their agent-relative information without being *about* the organism, or containing a self-referring component.

One might object that Perry is concerned with the (conceptual) content of belief and its truth conditions; the notion of an "unarticulated constituent" refers to an element that figures in the determination of the truth conditions of a belief without being explicitly represented. In contrast, what I am concerned with here is the (nonconceptual) content of perception and bodily experience itself, not the content of the beliefs that a subject can form based on her perceptions. I nevertheless take it that we can use Perry's insights to elucidate how we should think of the content of perception and bodily awareness. In the view proposed here, perception and proprioception provide the organism with information that is necessarily self-related and enables the organism to coordinate its perception and action to engage and interact with its environment. But precisely because the perceptual content is always related to one particular subject, the subject itself does not need to be identified or represented—similar to the linguistic case of the Z-landers, in which the location does not have to be represented because it always remains fixed. So the point I want to make is that the contents of perception and bodily awareness themselves do not contain a self-referring component—independent of whether a belief that is formed based on perceptual or bodily experience does.

Notice also that, in the view I sketched in chapter 2, nonconceptual content should not be characterized in terms of truth conditions. This raises the question as to how the notion of an "unarticulated constituent" can be transferred from belief to perception. As I have suggested, the nonconceptual content of perception provides the subject with a form of knowing-how (e.g., "knowing-how-to-interact-with-the-tree-in-front"), not with a form of knowing-that (e.g., "that there is a tree in front"). While the latter is truth evaluable, the former is not. However, the subject could, if she possessed the relevant concepts, form a belief on the basis of her perception (such as the belief "that there is a tree"), which would count as a true perceptual belief if and only if the subject of the perception does indeed stand in a perceptual relation to the tree. That is to say, the subject does play a role in determining the truth conditions of a belief that is formed based on her perception. But this is not to say that the subject (or the perceptual relation between the subject and the tree) must be represented in the perception—nor must the subject be represented in the content of the resulting perceptual belief. As Recanati (2007) puts it, it can figure as part of the "circumstance of evaluation"—as determined by the mode of experience—without being part of the content. Likewise, in the case of bodily experience, a belief that "legs are crossed" that is formed on the basis of the experience of one's crossed legs will be true if the property in question (that of crossed legs) is in fact instantiated in oneself. But that is not to say that either the self or the instantiation relation must be represented in bodily experience (cf. Recanati, 2009, pp. 266–267). Thus we can say that the self is an "unarticulated constituent" of the content of perception and bodily experience because the judgments that are formed on the basis of this content will have to be truth-evaluated with respect to the subject, although the subject herself (and the perceptual or instantiation relation) is not represented in perception and bodily experience—and neither does she have to be represented in the corresponding judgment, though on occasion she might be. So in a sense the notion of the self as an "unarticulated constituent" of perception and bodily experience is derivative upon the role that the self plays in evaluating the relevant beliefs that the subject could form on the basis of her perceptual states and bodily experiences.

Finally, let me emphasize that a being who possesses the first-person concept could, of course, on the basis of having a perceptual experience of a tree in front, not only form the belief that there is a tree, but also form the first-person judgment "I see a tree," thereby explicitly self-ascribing this experience. (In fact, a subject who possesses the first-person concept not only could but would also be justified in forming this judgment.) That is to say, the subject can, but does not have to, be represented in the content of a (conceptual) belief that is formed based on the (nonconceptual) content of her perception or proprioception. But the fact that perceptual content can, in the case of beings who possess the first-person concept, provide a subject with a basis for forming this explicit judgment neither implies nor requires that the perceptual content itself must contain an explicitly self-referring component. And it is only the content of perceptual (or bodily) experience that we are concerned with here. I return to this point in the next chapter.

What this discussion shows is that one can have self-related information without any self-representation. If so, it is more parsimonious to assume that the self is not represented in perception and bodily experience. Perception and bodily experience employ "monadic notions" rather than "relational notions," to use Campbell's terminology; there is no explicit self-reference. Hence the subject cannot be aware of referring to herself when being in a relevant experiential state. But the latter—awareness of the self-referring nature of a thought or experience—was supposed to be a necessary condition for "I"-thoughts or *de se* thoughts, and hence for self-consciousness. Recall our analysis from chapter 1. Self-consciousness was defined as the ability to think "I"-thoughts, where "I" thoughts are such that they nonaccidentally refer to the subject entertaining them. That is to say, in having an "I"-thought, the subject cannot but be aware that the thought refers to herself. But not only does this require that the content of the thought is necessarily self-related; it must also involve a self-referring component, for how are we to make sense of a nonaccidentally self-referring thought that does not actually contain a self-reference? Accordingly, perception and bodily experience cannot be considered forms of self-consciousness in the sense of the ability to think "I"-thoughts. Ecological perception and bodily experience provide the subject with information that is in fact *concerning* herself, but since that experience does not contain any self-referring component, it is not presented to the subject as being *about* herself.

Notice that there are also phenomenological reasons for resisting the thought that every conscious experience is also self-conscious. For example, when I am engrossed in the perceptual experience of a painting in front of me, or focused on a bodily activity, such as riding my bike or playing football, I am not also aware of myself as such. I am focused on the picture I am looking at, on the trajectory of my bike, or on the ball I am trying to pass, but not on myself; there simply does not seem to be an element in my experience that corresponds to the self. In fact, if I were self-conscious while enjoying my perceptual experience of the painting or performing an activity, this self-awareness would be a distraction from the task at hand. So although conscious experience is always experience *for* a subject and is experienced from the perspective of the experiencing subject, this is not to say that the subject must also be aware of herself or of having this particular perspective.[10]

To summarize the argument so far: According to the analysis in chapter 1, self-consciousness can be defined as the ability to think "I"-thoughts. "I"-thoughts are defined such that a subject who entertains an "I"-thought must be aware that the thought refers to herself. But one can only be aware that a thought refers to oneself if the thought actually contains an explicitly self-referring component (i.e., if the thought is actually self-referring).[11] However, this is not the case in perception and bodily experience. Therefore they are not forms of self-consciousness. And in conclusion, theories of nonconceptual self-consciousness are incomplete. To be sure, such theories can account for the action-guiding nature of certain representational states, which is one important characteristic of "I"-thoughts, but they fail to establish that the representational states in question actually represent the self, rather than simply containing implicitly self-related or agent-relative information.[12] Moreover, self-representationalist theories of nonconceptual self-consciousness put an unnecessary cognitive burden on even relatively simple intentional organisms because they seem to imply that self-representation is necessary for any kind of intentional interaction with the world, when the same interaction could also be explained in the absence of self-representation.

The self-representationalist might object that, contrary to the foregoing arguments, the self does directly appear as an object in visual perception and somatic proprioception. For instance, with regard to the content of perceptual experience, Bermúdez argues that the self "has a place in the content of perceptual experience in virtue of the self-specifying information that is an integral part of that perceptual experience" (Bermúdez, 1998, p. 108). As we have seen, this information consists in (a) information about bodily invariants that bound the field of vision, (b) information from visual kinesthesis about the movement of the perceiver, and (c) information about the possibilities for action and reaction that the environment affords the perceiver (cf. Bermúdez, 1998, p. 114). However, in the view I am defending here, the structural invariants that make up the self-specifying information in question are just part of what constitutes the egocentric perspective of the subject. It is true that perception is necessarily perspectival; the world is always experienced from the subject's perspective. However, it is one thing to perceive the world from one's egocentric perspective, and quite another to be aware of one's perspective in first-person terms (cf. Baker, 1998; Kapitan, 2006).[13]

A related objection to the argument might be that since perception always contains information about the world as well as about the perceiving organism, there has to be a mechanism that is able to distinguish between self and non-self (Vosgerau, 2009). As we have seen in the previous sections, every perceptual input carries information about both the perceiving subject and the world. Accordingly, the organism must somehow divide the input into self-related and world-related information. For instance, Vosgerau (2009) proposes a model that is based on the comparator model we encountered in chapter 1 and on the notion of sensorimotor contingencies (O'Regan & Noë, 2002), which I discussed in chapter 2. According to Vosgerau, "Whenever there is a motor command (efference or efference copy), there will be a typical change in the sensational input that systematically covaries with the command. These contingencies can be detected such that for each motor command the typical change in the sensational input can be learned. Sensation can hence be divided into two classes: the class that is caused by the system itself and the class that is caused by the world" (Vosgerau, 2009, p. 102; for a slightly different proposal, see also Legrand, 2006). But notice that the point of my argument here was precisely that although perception and bodily experience necessarily involve implicitly self-related as well as world-related information, the self is not presented to the subject as part of the representational content of these mental states. Indeed, as Vosgerau (2009) himself acknowledges, while the mechanism involved in the self–world distinction might be able to explain how the self-related information is processed at a subpersonal level, this does not imply that the system represents the self at the personal level. And, as we saw in the previous chapter, when we are considering representational content in the context of intentional action explanation, we are concerned with a personal-level (or animal-level) explanation. Personal- or animal-level explanations aim at representational content that is accessible to the subject in question such that it can provide the basis for the kind of practical reasoning processes associated with intentional action. Subpersonal-level explanations, on the other hand, describe brain processes underlying personal-level representations. It is certainly the case that such subpersonal-level explanations can contribute—as enabling or making-possible mechanisms—to an understanding of personal-level phenomena (Hornsby, 2000; Hurley, 2008; McDowell, 1994). For instance, as we saw in chapter 1, breakdowns in efference copy mechanisms have been suggested as part of the explanation for the phenomenon of thought insertion in schizophrenia (Frith, 1992). Under certain circumstances, knowledge of such mechanisms may even enter into the constitutive conditions of personal-level phenomena (Colombo, 2013). Nonetheless it would be fallacious to simply project properties between the personal and subpersonal level (Hornsby, 2000; Hurley, 2008; McDowell, 1994; Millikan, 1993). Accordingly, although it makes perfect sense to postulate such a self–world distinction mechanism at the subpersonal level, the existence of such a mechanism does not undermine the argument presented earlier. Quite to the contrary, postulating a self–world distinction mechanism is quite compatible with the notion of the self as an "unarticulated constituent" because it can explain on the basis of which subpersonal mechanism(s) perception and bodily experience contain self-related information in the absence of explicit self-representation.

Moreover, one might wonder whether this argument equally applies to perception and bodily experience. After all, while it seems plausible that the content of perception really is just about the external objects of perception, bodily experience is about one's own body. As Bermúdez puts it: "What somatic proprioception offers is an awareness of the body as a spatially extended and bounded physical *object* that is distinctive in being responsive to the will" (Bermúdez, 1998, p. 150; italics mine). According to Bermúdez, the body appears as an object of proprioception, as proprioception provides the subject with an awareness of the boundaries of the body. Hence one might think that—provided that the self is embodied—bodily experience therefore really is about oneself. However, although bodily awareness provides the organism with information that is *de facto* about its body, and although this information is in fact about the organism's *own* body, and thus about the organism, the information is not being represented *as such*. That is to say, it is not being represented *as* being about one's own body (rather than anybody else's). The argument role of the subject in one's states of bodily experience always remains the same and hence does not have to be explicitly represented; it remains "unarticulated." The relation to the subject is provided by the *functional role* or *mode* of the experiential state, which is such that it necessarily represents the bodily states and properties of one's own body, not by its explicit representational content. As Kapitan puts it: "That direct proprioceptive awareness and other forms of inner awareness are exclusively of a *unique* center of reception and reaction … obviates the need for any separate first-person representation" (Kapitan, 2006, p. 400; italics in original). So although bodily experience delivers information that is in fact about the organism, this information is not being represented *as* being about the organism. In other words, if one were to specify the content of such a representation, it would be "legs are crossed" rather than "my legs are crossed," for example. Notice also that what the quote from Bermúdez illustrates quite nicely is that the self-representationalist continues to construe self-awareness along the lines of object awareness. I argued in chapter 1 that such an approach is deeply problematic; we will return to this point in the next chapter.

Finally, one might ask, isn't implicitly self-related information just the kind of notion of self-consciousness that the self-representationalist needs? Isn't it just a terminological issue whether we call this primitive self-consciousness or not? Of course, one can decide to call experiences that contain implicitly self-related information forms of primitive self-consciousness. However, the notion of implicitly self-related information does not capture what the self-representationalist means by primitive self-consciousness, for what the self-representationalist is after is not just implicitly self-related information. Rather, the self-representationalist argues that "somatic proprioception counts as a genuine form of self-consciousness" (Bermúdez, 1998, p. 144), which represents bodily properties "*as* properties of one's self" (p. 145; italics in original). However, this requires explicit self-representation. And, as I have just tried to show, the content of perception and bodily experience, while containing self-related information (and hence being *self-concerning*), does not explicitly represent the self.

The crucial upshot of all of this is that the self-representationalist seeks to establish a theory of nonconceptual self-consciousness based on self-representation, whereas in my view, we ought to focus not on the representational *content* of, for example, perception and bodily experience but rather on their *mode* of presentation if we want to explain how self-consciousness can emerge from nonconceptual forms of representation. (I develop this point in more detail in the next chapter.) This is a fundamental difference. It will become even clearer when we turn to one of the other criteria for genuine self-consciousness, namely, immunity to error through misidentification, and when we consider in more detail the structure of nonconceptual content.

### 3.4.2  Immunity to Error through Misidentification Reconsidered

As we saw earlier, proponents of theories of nonconceptual self-consciousness argue that perception and bodily experience are immune to error through misidentification, and this is an important criterion for "I"-thoughts. In this section, I argue that this view is mistaken. This is because, due to its noncompositional structure, the nonconceptual content of perception and bodily experience not only does not have to contain an explicitly self-referring component to be agent relative and thus action guiding but also cannot contain such a component. If so, the notion of immunity does not have a grip on perception and bodily experience.

Recall that nonconceptual content is defined negatively, in opposition to conceptual content. As was shown in chapter 2, nonconceptualists, in contrast to conceptualists, hold that there are ways of representing the world that do not require concepts. Moreover, members of the debate between conceptualism and nonconceptualism generally agree that conceptual content consists of several components that can be decomposed and recombined to allow for the context-independent systematicity and productivity of thought. In other words, conceptual content is generally considered to meet Evans's (1982) Generality Constraint. In contrast to conceptual content, nonconceptual content is thought to be essentially noncompositional and thus is not subject to the Generality Constraint (cf. Meeks, 2006; Heck, 2007; Hanna, 2008; Toribio, 2008). We might also say that it is structure implicit, which is to say that there is no explicit representation of object, predicate, or predication relation. The nonconceptual content of experience "provides a way of carving up the world which is not a way of carving it up into object, properties, or situations (i.e., the components of truth conditions)" (Cussins, 2003, p. 134). Rather, as was discussed in chapter 2, nonconceptual content is to be spelled out in terms of the organism's skills and abilities to move and interact with the world, and so in terms of the affordances provided by the environment.

Importantly, this means that nonconceptual content does not have the necessary structural features required for an explicit predication relation of the form "a is F" (which could be decomposed into object and predicate) or, for that matter, "I am F," and hence does not have the necessary structural features required for an explicit self-representation—as in "I am hungry," "I am in pain," "I see a tree in front of me," and so on. The structure required for explicit self-representation can only come with content that is compositional, and thus with conceptual content. So far, this argument lends further support to the claim established in the previous section, namely, that there is no explicit self-representation in perception and proprioception (although the argument of this section is based on the noncompositionality of nonconceptual content and not on the fact that self-representation is not necessary for perception and action guidance).

However, that nonconceptual content is thought to be noncompositional also implies that nonconceptual content cannot be said to be immune to error through misidentification. Let us recall the basic definition of immunity to error through misidentification. As was shown in chapter 1, a thought that is immune to error through misidentification allows for the possibility that it misrepresents the property that is being ascribed, while it cannot misrepresent the subject purportedly possessing that property. The reason for this is that thoughts that are immune to error through misidentification are based on ways of coming to know about the world such that it is impossible that I come to be aware of a certain property, but that the person having the property is not me. For instance, consider the property of being hungry. Frequently, I start feeling irritable when I am hungry, thereby arguably misrepresenting my state of hunger for a state of irritation. However, if, on the basis of my experiential state, I (mistakenly) self-ascribe the state of being irritated, then it is impossible for me to misidentify the subject of this state. While it is possible (or even likely) that I am mistaken with respect to the property I am self-ascribing, I cannot be mistaken with respect to the subject of the property in question. In other words, such a self-ascription can "what"-misrepresent, but it cannot "who"-misrepresent.

What this means, however, is that the notion of immunity is applicable only to judgments containing an explicit "who"-component.[14] This, in turn, requires conceptual representations, for the notion of a representational distinction between subject and property, which makes the idea of "what"-misrepresenting rather than "who"-misrepresenting a sensible one (or vice versa), is a matter in the realm of conceptual contents.[15] Nonconceptual content does not "what"-represent *and* "who"-represent. Rather, the representation is in a format in which the "what" is necessarily implicitly self-related, and the "who" is not represented at all. But if nonconceptual content does not "what"- *and* "who"-represent, then it does not make sense to ask whether it can misrepresent the "what" while accurately representing the "who." This means that the question of whether nonconceptual content is immune to error through misidentification cannot even arise. States with nonconceptual content can neither be immune nor subject to error through misidentification. In other words, the question of whether nonconceptual content can be immune to error through misidentification is ill posed because the question can only arise at the level of *judgment*, not at the level of nonconceptual content. It is a category mistake, so to speak, to try to apply the notion of immunity at the level of nonconceptual content.

Notice that *if* the self-representationalist was right that the self is represented in the nonconceptual content of experience, and *if* it made sense to apply the immunity principle to states with nonconceptual content, these states would still fail to be immune to error through misidentification. This is because, as Meeks (2006) has argued, nonconceptual content cannot be immune to error through misidentification because it is noncompositional. Meeks argues that, to count as representational content, nonconceptual content must allow for the possibility of misrepresentation. Further, because states with nonconceptual content are noncompositional, they lack the internal structure required to distinguish between subject and predicate. But this implies that states with nonconceptual content do not allow for a misrepresentation of the property while at the same time accurately representing the subject of the state. In other words, according to Meeks, insofar as nonconceptual content is noncompositional, if there is to be any possibility of misrepresentation at all, it would have to be a misrepresentation *tout court*. That is, it would misrepresent the subject along with the property that is being represented. Conversely, to say that a state with nonconceptual content is immune to error through misidentification is to say that it cannot misrepresent at all; but this is incompatible with the notion of it being a representational state. In conclusion, according to Meeks, states with nonconceptual content cannot be immune to error through misidentification.

Notice how Meeks's argument differs from the argument just given. Again, Meeks's argument would hold *if* the self-representationalist was right that the self is part of the explicit representational content of experience, and *if* the immunity principle could be applied to states with nonconceptual content. However, I have just argued that both of these assumptions should be rejected. First, since the notion of immunity can only apply to judgments with a subject–predicate structure, it makes no sense to ask this question when considering states that lack such a structure. Second, Meeks's argument implies that insofar as states with nonconceptual content can misrepresent (which is indeed a necessary requirement for them to count as representational states), they would equally misrepresent the subject and the predicate, as these are necessarily intertwined. In contrast, I have just argued that the subject is not part of the explicit representational content at all; hence it cannot be misrepresented, even if the predicate in question is misrepresented. In fact, it is precisely because of this that perceptual states with nonconceptual content can provide the *foundation* for first-person judgments that are immune to error through misidentification (even if they themselves do not fall into this category). I discuss this point in more detail in the next chapter. In sum, while Meeks argues that states with nonconceptual content must be subject to error through misidentification, my argument here shows that they are neither immune nor subject to error through misidentification.

Either way, this poses a problem for the proponents of nonconceptual self-consciousness, since they are no longer entitled to claim that the nonconceptual content of perception or bodily experience is immune to error through misidentification. It follows that perception and bodily experience should not be considered forms of nonconceptual self-consciousness. One of the main arguments in favor of the view that ecological perception and bodily experience are to be considered nonconceptual forms of genuine self-consciousness is that they are immune to error through misidentification. However, as I have just argued, the notion of immunity does not apply to nonconceptual content; hence the argument put forward by the self-representationalist loses its force.

Again, this is not to say that we can in fact be mistaken about who the subject of our perceptual or proprioceptive states is. It would be equally wrong to say that perception and bodily experience are subject to error through misidentification. The fact remains that ecological perception and bodily experience contain information that is self-related, and that this is necessarily so. Rather, because this information is represented in a way in which the subject of perception is not a component of the explicit representational content, we cannot say that there is a possibility of "what"-misrepresentation while leaving the representational "who" intact—as there simply is no representational "who" that could be represented independently of the representational "what." But the possibility of explicitly representing the subject independently of the specific properties that are ascribed to the subject—at least in principle—is a necessary precondition for the notion of immunity to gain any traction. Therefore ecological perception and bodily experience can at best be said to provide the *basis* for the formation of first-person judgments that are immune to error through misidentification and thus constitute paradigmatic forms of self-consciousness, but they do not by themselves constitute such paradigmatic forms of self-consciousness.

Three objections might be brought forward. First, by defending the so-called state view rather than the content view of nonconceptualism, one could evade the problems raised here. As we have seen in chapter 2, for the state nonconceptualist, it is consistent to hold that all content—whether perceptual or belief content—is conceptual, but for a subject to undergo a perceptual experience, she does not need to possess the concepts required to specify the content of her experience. Accordingly, it would be consistent for the state nonconceptualist to hold that a state with nonconceptual content can, contrary to the argument just given, represent the subject of the state independently from the property, and it can misrepresent the property without misrepresenting the subject to which the property is being ascribed. It is just that the subject being in a state with that kind of content does not need to possess the concepts necessary to spell this out. However, as I argued in my discussion of the state view in chapter 2, the state view does not represent a coherent version of nonconceptualism. To remind ourselves why this is so: The motivation for introducing a different notion of content in the first place was that it helps us understand the ways in which subjects can grasp the world as being, and the idea is that not all these ways require concepts—either because the subject in question (e.g., an animal or a prelinguistic infant) does not possess them, or because the representational state in question does not require them, as is arguably the case with certain types of perceptual experience. But, as Toribio (2008) has argued, either the state view entails a notion of content that is unable to capture the way a subject grasps the world as being (because this notion is divorced from the cognitive abilities of the subject), and hence cannot help us explain the subject's intentional behavior, thus undermining the justification for introducing the notion in the first place; or it entails that a subject could exercise cognitive abilities she does not possess, which is incoherent. Thus, adopting the state view of nonconceptualism is not a feasible option.

Second, one might object that things are not quite as simple as how they are being presented here, for nonconceptual content could fail to satisfy Evans's Generality Constraint without being entirely noncompositional.

Indeed, Peacocke (1992, chap. 3) argues explicitly in favor of what he calls proto-propositional content at the nonconceptual level, and Bermúdez (1998, chap. 3) likewise argues that nonconceptual content is compositional, and that we can ascribe "proto-beliefs" to nonconceptual creatures. There is a sense in which this objection hints at an important point: namely, that the distinction between conceptual and nonconceptual content is not fine grained enough to capture all the different levels of representation that can be found between relatively simple perceptual states, which can be characterized as simple feature-placing states, and states with full-fledged conceptual content (see chap. 2). Indeed, it is precisely because the simple dichotomy between conceptual and nonconceptual representations does not do justice to the different levels of representational complexity and flexibility that I will argue in chapter 5 that we need an account that goes beyond this simple dichotomy. That is to say, I will argue for a gradual transition between nonconceptual and conceptual representations. Accordingly, there will be states with content that displays a partial grasp of certain concepts, including the self-concept. As I will argue in chapter 5, the distinction between implicit and explicit representation (and in particular between different levels of explicitness) can be used to account for these different levels of representation. If so, it is indeed conceivable that there could be representational states that allow for a certain level of explicit self-representation while only implying a partial possession of the relevant concepts, and that these states could be classified as being immune to error through misidentification (see chap. 6). Nonetheless I take it that this does not apply to simple perceptual or bodily experiences, which have been the focus of our discussion so far.

This is not just because I take it that perceptual and bodily experience do not possess the required degree of complexity. Rather, the objection misses its target because, again, with respect to perceptual and bodily experience, the main issue is not in fact compositionality. Even if we were to grant a certain level of compositionality to these nonconceptual representations, doing so would not get around the problem that I discussed earlier, namely, that the contents of perception and bodily experience do not contain a self-referring component. That is to say, there simply is no "who"-representation in perception and bodily experience. The content of perception and bodily experience does not contain a component that could stand for the "I" of a self-conscious thought. And since the immunity principle can only apply to representational states that can at least in principle provide for a "who"-representation that is independent from any specific "what"-representation—and would be articulated by means of the first person—the immunity principle cannot apply to states that lack such a self-referring component. So the immunity principle fails to apply to so-called nonconceptual forms of self-consciousness such as perception and bodily experience because no self-referring component is needed to account for the essential self-relatedness and action-guiding function of perception and bodily experience.

A third objection might be that the argument is not being fair to proponents of self-representationalist accounts because it is setting up a straw man. Perhaps a charitable reading of the self-representationalist account would commit it not to the claim that perception and bodily experience are themselves immune to error through misidentification, but rather to the claim that they *ground* judgments that are immune to error through misidentification. This would be perfectly consistent with—indeed, it would even be an outcome of—the argument just given. And perhaps this really is all the proponent of such a theory needs to establish. However, notice first that Bermúdez states explicitly that "the fact that somatic proprioception *is itself immune to error through misidentification relative to the first-person pronoun* provides vital support for the claim that somatic proprioception counts as a genuine form of self-consciousness" (Bermúdez, 1998, p. 144; italics mine). It follows that if the notion of immunity to error through misidentification fails to have a grip on the nonconceptual content of bodily experience, then bodily experience cannot count as a genuine form of self-consciousness. Yet, perhaps Bermúdez is simply misspeaking at this point. Nevertheless, even if all that was to be shown is that perception and bodily awareness can provide the *foundation* for first-person judgments that are immune to error through misidentification, the way to go about showing this is not by arguing that the self is represented in perception and bodily awareness. To the contrary, it is only once we accept that the self is *not* represented in perception and bodily experience that we can give a proper account of immunity to error through misidentification, as well as of the subjectivity of conscious experience (see next chapter). Hence, while I agree that we should—and can—show how perception and bodily experience can ground first-person judgments that are immune to error through misidentification, we should reject the self-representationalist account because it chooses the wrong strategy for showing this. Instead we should opt for a "no-self" account, which I spell out in the next chapter.

## 3.5 Conclusion

The conclusion to be drawn from this chapter is that, rather than trying to solve the problem of self-consciousness by giving an account of nonconceptual self-representation, we should aim for a non-self-representationalist, or "no-self" account. So rather than trying to explain the possibility of conceptual self-reference in terms of nonconceptual self-reference—that is, in terms of the *content* of experience—we should explain the possibility of conceptual self-reference in terms of the *mode* of experience. A "no-self" account not only does justice to the arguments against the belief that the self is represented in, say, perception and bodily experience but, as I show in the next chapter, is also better able to account for the phenomenon of immunity to error through misidentification. This is in part because it constitutes a radical departure from the problematic subject–object model of self-consciousness, while self-representationalist theories remain implicitly committed to this model. Consequently, as we will see in the next chapter, it is able to provide an analysis—albeit a deflationary one—of the "sense of mineness," and it might also provide a way of interpreting Hume's famous elusiveness thesis (see chap. 1). Moreover, it has interesting implications for an account of the possession conditions of the first-person concept. So the view I defend in what follows is a non-self-representationalist view.

I take it that the account I intend to propose will be compatible with other recently developed accounts in terms of the rejection of self-representationalism (e.g., O'Brien, 2007; Peacocke, 1999; Recanati, 2007; see also Moran 2001 & Rödl, 2007). However, my account will also extend significantly beyond these other accounts because while they are able to do justice to the distinction between implicitly self-related information and explicit self-representation, they do not explain how we get from the former to the latter. In the remainder of the book, I aim to address this gap. In particular, in chapters 5 and 6, I develop an account of implicit and explicit representation that distinguishes different levels of explicitness. This will allow for a more fine-grained analysis of the precursors to and levels of self-conscious thought and will thus also lead to an improved model of the development of self-consciousness. So the model I propose will complement the insights provided by theories of nonconceptual self-consciousness—and in particular by non-self-representationalist accounts—with an analysis of the transition from implicit to explicit self-representation. As we will see in chapter

6, this will also require an investigation of the relationship between self-consciousness and intersubjectivity, a relation that has been neglected by many theories of self-consciousness (with notable exceptions).

## 3.6   Coda: On the Neuroscience of Self-Consciousness

It is an interesting question whether and how the arguments presented here relate to recent developments in the neuroscientific study of self-awareness. Although, as mentioned earlier, it is important to distinguish between personal and subpersonal level explanations, assuming that the personal supervenes on the subpersonal, we should expect differences at the personal level to be reflected at the subpersonal level, and subpersonal-level explanations can contribute—as enabling or making-possible mechanisms—to an understanding of personal-level phenomena. Over the past few years, neuroscientists have begun to study the neural correlates of what has been termed "self-referential processing" (for a recent review, see Qin & Northoff, 2011). Self-referential processing is the processing of stimuli that a subject judges to be relevant to herself (such as her name, a picture of herself, or certain character traits that she ascribes to herself). The neural correlates of this kind of processing have been identified with structures in the cortical midline structures of the brain (Northoff & Bermpohl, 2004). Interestingly, a strong overlap seems to exist between the structures associated with self-referential processing and the structures associated with the resting-state activity of the brain (e.g., d'Argembeau et al., 2005; Schneider et al., 2008)—the so-called "default-mode network" (Raichle et al., 2001). Neural activity in these structures has been found to be inhibited during demanding perceptual tasks, whereas increased activity is thought to be associated with self-reflection and mental time travel (e.g., Goldberg et al., 2006). This seems to support the view that explicit self-representation is to be distinguished from the processing of self-related information that is implicit in, for example, perceptual experience.

However, there are also important caveats against any such conclusion. It is questionable whether the paradigms employed in neuroscientific studies of self-referential processing and its relation to the default network reflect the distinctions that are crucial to the present discussion. After all, proponents of the self-representationalist view would agree that there is a difference between the kinds of judgments involved in evaluating stimuli as being self-relevant or not and what they take to be nonconceptual forms of self-consciousness (e.g., perception and bodily experience), so these studies are not suited to distinguishing between the self-representationalist and the non-self-representationalist view as such. (For a more detailed critical discussion, see also Musholt, 2013b.) Moreover, it is unclear what exactly is going on during the so-called resting state. Resting-state activity is the kind of activity that occurs in the absence of any external stimuli. However, we do not know what exactly subjects are doing during these periods; possible activities include the processing of internal stimuli, mental time travel, explicit self-reflection, and others.

What might be more promising in terms of establishing a link between neuroscientific approaches to self-consciousness and the view proposed here is Northoff's distinction between "self-referential" and "self-related" processing (see Northoff, 2011, 2013, esp. vol. 2). Self-referential processing refers to the explicit judgment or awareness of the self-specificity of a stimulus, while self-related processing refers to the general processing of any stimulus in relation to the resting-state activity of the brain, independent of the subject's awareness of whether the stimulus is relevant to herself. However, although potentially promising, the notion of self-related processing as defined by Northoff seems to be somewhat different from the way it is standardly used in the scientific literature (Christoff et al., 2011). It also remains unclear how exactly it compares to the notion of implicitly self-related information that was the subject of discussion in the preceding paragraphs. On the one hand, in contrast to the way in which the notion was discussed here, the notion as employed by Northoff seems to refer to a characterization of basic neural coding, prior to and independent from sensorimotor processing, as well as from the existence of an egocentric perspective on the world. In contrast, I have argued that the essential self-relatedness of perception, say, is a result of the fact that perception is always from the perspective of the subject. On the other hand, in Northoff's view, self-related processing also seems to be closely associated with the activity of the default network, which, as we just saw, is thought to be involved in diverse functions related to self-*reflection*. (For further criticism of this notion and its philosophical relevance, see also Christoff et al., 2011.) Therefore, while the neuroscientific study of self-consciousness has produced some interesting results, we cannot yet draw any firm conclusions from these results with respect to the arguments defended here.

# 4  Toward a Non-Self-Representationalist Account

In the account defended here, self-consciousness, that is the ability for self-referential, first-personal thought should not be seen as being grounded in forms of nonconceptual self-representation. This means that the first-person concept does not refer to a representational *object*, namely, the self, that is specified in terms of particular types of information, such as those described by Bermúdez. This is not to say that the first-person concept does not refer at all; the first-person concept does refer to the *subject* of the thought in question. But the way in which the first-person concept refers is not by means of exploiting the self-representational *content* of conscious states but rather by means of exploiting a specific *mode* of presentation.

Indeed, one of the problems of the self-representationalist account is that it remains implicitly committed to the subject–object model of self-consciousness. That the self is thought to be part of the explicit representational content of, say, perception and bodily awareness implies that self-consciousness is still being construed as the self being presented to itself in experience, and hence is still being construed along the lines of object awareness in this account. In the self-representationalist account, perception and bodily experience deliver information *about* the self; that is, the self remains an *object* of intentional experience, albeit in a way that makes the self the only possible intentional object. So, for example, the content of perception in this model would have the following structure (cf. O'Brien, 2012, p. 143):

(a)  $x$ perceives $x$ to be standing in front of a tree.

In this view, the content of perception contains an explicitly self-referring component, that is, an argument role that stands for the subject of perception. In contrast, we have seen in chapter 1 that self-consciousness should not be construed along the lines of object cognition. Moreover, I argued in the previous chapter that although the contents of perception and bodily experience are self-related, they do not represent the self; that is, they do not contain a self-referring component. This is precisely because the information contained in these states is gained in ways that are specific to the self; we cannot gain information about others in the same way. Such a self-specific way of gaining information can be called an "internal mode" (Recanati, 2007). And because the information that is gained by means of such internal modes of presentation is necessarily self-related, the self does not need to be explicitly represented in the content of the resulting experiential states. For example, I cannot become aware, from the inside—that is, through bodily experience—that someone else's legs are crossed; information gained by means of bodily experience is necessarily self-related. Importantly, the essential self-relatedness is provided by a fact that is *external* to the content of the experience itself, namely, by the mode of presentation— or, put differently, by facts about my cognitive architecture. So in the non-self-representationalist account, the structure of perception should rather be construed as follows:

(b)    $x$ is aware of standing in front of a tree.

As we will see in the following, such a "self-less" view of the structure of experience can both help us provide an account of the sense of "mineness," or the subjectivity of experience, while at the same time doing justice to Hume's insight, and explain the fact that judgments that are formed based on this experience are immune to error through misidentification.

The chapter proceeds by demonstrating how the non-self-representationalist view can account for the phenomenon of immunity to error through misidentification (sec. 4.1). This section also discusses how the view I am defending here relates to the views of Recanati (2007, 2012) and Peacocke (1999, 2012), and considers a potential objection based on Peacocke's (2012) notion of nonconceptual *de se* content. This leads to an analysis of prereflective self-consciousness in terms of the "no-self" view (sec. 4.2). Section 4.3 revisits the issue of immunity and considers a potential objection to my view on the grounds that it cannot account for logical immunity.

## 4.1   Selflessness and Immunity to Error through Misidentification

Let us begin by examining how the non-self-representationalist account can provide a way of accounting for the phenomenon of immunity to error through misidentification. In the self-representationalist view, we have to assume that the self is represented in experience, albeit in such a way that the self is the only possible object being so represented. However, it is hard to see why the self could not be misrepresented in the content of experience in this view. That is, it is hard to see why such self-representations should be immune to error through misidentification. Insofar as the self is represented explicitly in experience, there should, at least in principle, be the possibility of misrepresentation. Moreover, as we saw in the previous chapter, on the assumption that nonconceptual content is noncompositional, any possibility of misrepresentation would necessarily entail the possibility of misrepresenting the subject of the state in question in the self-representationalist view (Meeks, 2006).

In contrast, it is easy to see how an explicit first-personal judgment that is grounded in experiential states—or in an internal mode of presentation— can be immune to error through misidentification in the non-self-representationalist view. The reason for this is that such a judgment will not be based on a representation of the self; accordingly, there is no possibility of misrepresenting the self. In other words, as the self is not represented in the content of perception and bodily experience, perception and bodily experience do not present the subject with self-identifying information on the basis of which it could self-ascribe the experiential state in question. But if no such self-identifying information is made available, and if accordingly the resulting judgment is not based on a self-identifying process, there is no room for misidentification. Instead, a first-person judgment that is based on perceptual or bodily experience exploits the fact that the relevant mode of experience is self-specific by making this fact—which is implicit in the experience—explicit. Put differently, a subject can form a first-person judgment that is immune to error through misidentification by making explicit the implicit self-relatedness of the information contained in perception and proprioception via the application of the self-concept. Such a judgment will take the form "I see a tree in front of me" or "my legs are crossed," for example. It is important to note that such a judgment does not rely on any sort of identity judgment or inference. Accordingly, it does not introduce any room for error based on misidentification. Rather, it simply makes explicit what was already implicitly contained in the underlying experiential state that grounds the judgment, namely, the self-relatedness of the mode of experience. So it is the essential self-relatedness of internal modes of presentation that ultimately accounts for the immunity to error through misidentification of certain first-person judgments, provided that in the transition from implicitly self-related information to explicit (conceptual) self-representation no identity judgment is required (cf. Recanati, 2007, 2012).

That no such judgment is introduced in the formation of these judgments might be elucidated in terms of Peacocke's (1999) notion of a "primitively compelling judgment" (cf. Dokic, 2003). A primitively compelling judgment in Peacocke's sense is a judgment that a subject both can and should make without any further reasoning or justification. We are now in a position to better understand what this might mean in the context of self-consciousness: no further justification or evidence is required for the formation of certain "I"-thoughts (i.e., those where the "I" is used "as subject") because the judgment only makes explicit what was already implicit in the experiential state based on which the judgment is made.[1] For the present discussion, this means that the subject does not need any additional premise—and in particular no identity premise—when ascribing to herself, say, a bodily property, such as an itch in her right foot, based on the appropriate bodily experience (i.e., the feeling of itchiness), because the self-ascription only makes explicit the essential self-relatedness that was already implicit in the mode of experience. Thus this account not only does justice to the arguments presented in the previous chapter but also is better able to provide an account of immunity to error through misidentification.

The account presented here is closely related to Recanati's (2007, 2012) account of immunity to error through misidentification. Similar to the view defended here, according to Recanati, experiential states such as those involved in perception and bodily experience are self-less in the sense that they do not represent the self. Rather, the specific relation that the content of these states bears to the subject is provided by the experiential mode. Indeed, in his book *Perspectival Thought*, Recanati (2007) argued that immunity to error through misidentification can only be ascribed to these experiential states themselves, which he calls implicit *de se* thoughts. This is precisely because they do not represent the self and hence cannot misrepresent it. However, in his 2012 paper, he revised this view—and rightly so—by allowing that explicit first-person judgments can also be immune to error through misidentification, provided that they are grounded in the right (i.e., self-less, but necessarily self-related) experiential states. So in Recanati's revised view, such first-person judgments *retain* the immunity to error of the underlying mental state despite introducing an explicitly self-referring component—and hence amounting to an explicit *de se* thought—insofar as no additional evidence base (i.e., no extra identity premise) is introduced when the judgment is based on an internal mode of experience. After all, the judgment only makes explicit what was already implicit in the mode. While I agree that this revision was necessary, I would go even further and argue that *only* explicit first-person judgments can properly be called immune to error through misidentification. If the arguments that I presented in the previous chapter are correct, it is somewhat misleading to call experiential states such as those involved in perception and bodily experience "implicit *de se* thoughts," because it is difficult to understand how we can make sense of a *de se* thought (or "I"-thought) that does not involve self-reference. For the same reason, it does not make sense to call these states immune to error through misidentification. Instead, only once we move to the explicit self-ascription of the state in question does the notion of immunity gain traction. However, as far as the substantial issues are concerned, I agree with Recanati that the explanation for the possibility of "I"-thoughts that are immune to error through misidentification is to be found in the mode of presentation of experiential states and not in the fact that they represent the self.

I take it that the view proposed here could also be interpreted in terms of Peacocke's (1999) notion of "representationally free" uses of "I." According to Peacocke, "I"-thoughts (i.e., explicit self-ascriptions) that are made on the basis of conscious states or activities are "representationally free" uses of "I," which renders these self-ascriptions immune to error through misidentification.[2] Such self-ascriptions are characterized by the fact that "the thinker's reason for self-applying the predicate $F$ is *not* that one of his conscious states has the content 'I am $F$.' His reason is not given by the 'representation distinguishing a particular object' as $F$. It is rather the occurrence of a certain kind of conscious state itself which is his reason for making the judgment" (Peacocke, 1999, p. 285; italics in original). So such "I"-thoughts are not made on the basis of the fact that the conscious state in question represents the self; the self is not given as an object of experience. Rather, it is because a conscious state necessarily requires a subject, and because a subject can only ever experience its own mental state and not that of another, that a thinker who possesses the first-person concept is entitled to self-ascribe it. Possession of the self-concept then expresses recognition of, or rational sensitivity to this fact of ownership, a fact that is, as we have seen earlier, *external* to the experiential content. According to the arguments given so far, "I"-thoughts that are made based on visual or bodily experience should fall under this characterization.

However, in his 2012 paper, Peacocke claims that we ought to ascribe *de se* content to perception and action (contrasting his own view with that of Perry [2002]). This might be read as a potential objection to the view defended here, for I have been arguing that the self is not part of the content of such experiential states. However, it seems to me that the conflict between Peacocke's view and mine—if there is any—is of a superficial (i.e., terminological) rather than a substantial nature, and both views (i.e., Perry's and Peacocke's views, as well as Peacocke's view and the view proposed here) can be reconciled if we take into account the role of the mode of experience in grounding first-person judgments. Although Peacocke emphasizes the need for a nonconceptual *de se* notion, he takes this to be "an experience-independent indexical notion" rather than a "perception-based or a sensation-based demonstrative" (Peacocke, 2012, p. 155). Now, insofar as such a notion is based precisely on something other than self-representation, I hold that it can be accounted for by the role of the mode of experience, which, for the experiences in question, is such that it is necessarily self-specific. Possessing the first-person concept would entail sensitivity to this fact, which in turn would enable the self-ascription of the experience in question by means of applying the first-person concept. I see no reason why we need to postulate a nonconceptual self-notion to make sense of this. However, insofar as such a nonconceptual self-notion is thought to be non-self-representational and experience independent, it seems to me that the best way to interpret it is in terms of the contribution of the mode to the subjectivity of experience (see next section), in which case the disagreement would amount to nothing more than a terminological issue.[3]

## 4.2 Non-Self-Representationalism and the Sense of "Mineness"

Not only can the non-self-representationalist view make intelligible the immunity to error through misidentification of certain first-person judgments—namely, those that are grounded in internal modes of experience—but it might also help us to analyze what some philosophers call prereflective self-consciousness, or the sense of "mineness." As was shown in chapter 1, some authors claim that experience is necessarily endowed with a sense of "mineness," which is thought to be necessary to enable the subject to ascribe an experience to herself without having to rely on identification or self-observation. Such self-ascriptions are then thought to result in judgments that are immune to error through misidentification. Another way of putting this is by claiming that experience is necessarily "subjective," that is, experience is always experience for a subject. We also saw in chapter 1 that some authors have argued that the notions of pre-reflective self-consciousness or the sense of "mineness" have to be taken as primitive and cannot be further analyzed. But why not take the non-self-representationalist account to provide precisely what is claimed to be impossible according to these authors, that is, an analysis of the sense of "mineness"? In the view proposed here, judgments that are made based on internal modes of experience take the self "as subject" precisely because they are not based on content that represents the self. Given that the self is not represented as an object in the content of experience—for it is implicit in the mode—the only way in which it can be conceived in a judgment that is based on such a non-self-representational content is as the *subject* of experience (cf. Recanati, 2007, p. 194). In contrast, as I argued earlier, in the self-representationalist view, self-consciousness is implicitly construed along the lines of object awareness; accordingly, this view has difficulty accounting for the use of "I" "as subject."[4]

Importantly, this shows that it is not necessary to postulate a sense of "mineness" above and beyond the content and mode of experience. That is to say, the non-self-representationalist view provides us with what might be called a deflationary account of the sense of "mineness." In this account, the subjectivity of experience, or the sense of "mineness," is to be seen as a result of the combination of the representational content of experience, which presents us with intentional objects relative to our possibilities of interacting with them, and the mode of experience, which is specific to the self (i.e., perspectival and self-related). Experience is necessarily subjective, in the sense of being *for* a subject, because it is given in a mode that is specific to the experiencing subject; nothing else is required.

In fact, such an analysis turns out to be quite compatible with some phenomenological analyses, as well. This is because many phenomenologists seem to suggest that prereflective self-consciousness, or the sense of "mineness," is not to be understood in terms of self-representation.[5] After all, they often stress that the self does not appear as an object in prereflective self-consciousness (cf. Legrand, 2007; Poellner, 2003; Zahavi, 2005). For instance, Legrand claims that it is possible to be prereflectively self-conscious without representing the self (see Legrand, 2007, p. 591). Moreover, Zahavi, for example, also seems to think that prereflective self-consciousness is tied to the mode rather than the content of experience when he claims, "If the experience is given in a first-personal *mode of presentation*, it is experienced as my experience, otherwise not" (Zahavi, 2005, p. 124; italics mine). Indeed, in a forthcoming essay, Zahavi and Kriegel state explicitly that the view that experience necessarily comes with a sense of "mineness" does not entail that in addition to the objects that are being perceived, the self is also given as an object in experience; rather, the claim is that the objects of experience are presented to the subject in a distinctly first-personal way. So on their view, the sense of "mineness" does not refer to *what* is experienced, but rather to *how* things are experienced (see Zahavi & Kriegel, forthcoming). This seems to be compatible with my view, namely, that it is to the specific (i.e., internal) modes of experience, rather than to the content of experience, that we need to appeal so as to account for both the subjectivity of experience and the fact that such experiences can ground first-personal judgments that are immune to error through misidentification. It is by considering the mode of experience that we can explain how the self can in some sense be part of the experience—for in this view, experience is necessarily perspectival, which is to say *for* a subject—even though it is not represented in the content of experience.[6]

Notice that the non-self-representationalist view is also able to do justice—and perhaps illuminate—Hume's claim that, when engaging in introspection, we are always only aware of different experiential states, but never of the self as such (see chap. 1). One way of understanding Hume's thesis is that we never find the self in introspection because the self is, in fact, never represented in experience. Experience presents us with access to objects in the world, but the self as such is not among the objects that experience presents us with.[7]

### 4.3 Immunity Reconsidered: Which Kind of Immunity?

A potential objection to the account presented here is that it conflates different notions of immunity to error through misidentification. Readers who are familiar with the debate about immunity will know that it is common to distinguish between circumstantial and absolute immunity, on the one hand, and *de facto* and logical immunity, on the other. It might be argued that these distinctions are relevant for a theory of self-consciousness—for instance, because circumstantial or *de facto* immunity is not sufficient to ground self-consciousness (cf. O'Brien, 2007)—and that the foregoing discussion does not do adequate justice to these distinctions.

With respect to the distinction between circumstantial and absolute immunity, Recanati (2007) has rightly pointed out that this distinction seems to be misguided. Shoemaker (1968) thinks that first-person judgments involving the self-ascription of mental states, such as "I see a canary," possess absolute immunity, because the subject cannot be mistaken that it is she* who is having the visual experience in question. In contrast, a judgment such as "I am facing a table" is only circumstantially immune: it will be immune when it is made based on seeing the table in front of oneself, but it will not be immune if it is made based on, for example, seeing oneself in the mirror, for in this case it is possible to misidentify the person one sees in the mirror. Similarly, it would seem that the judgment "my legs are crossed" is immune when made based on bodily experience, but not when seeing one's legs in the mirror. The problem with this view, as Recanati points out, is that immunity "is *always* circumstantial, that is, always relative to a 'way of gaining information' (Evans, 1982)" (Recanati, 2007, p. 149; italics in original). The information that I am facing a table or that my legs are crossed, as well as the information that I see a canary, can be gained in ways that do not give rise to immunity.

For example, imagine I am a world-famous neuroscientist suffering from blindsight and undergoing a typical blindsight experiment. Images are shown to me, which I am supposed to identify. Presently an image of a canary is displayed but, being a blindsight patient, I am not aware of seeing a canary (even though I can be shown to have—subpersonally—identified the canary). At the same time, however, the electric activity of some of my neurons involved in visual identification is being recorded and amplified online in the form of a crackling sound, which I can hear. Thanks to an elaborate theory of mine, I am able (or think I am able) to identify what I see from the sound the neurons make. In such a situation I may assert "I see a canary" because I hear what I take to be the typical sound of canary-neurons. My assertion is clearly not immune to misidentification: for, unbeknown to me, I might well be listening to the neurons of some other patient undergoing the experiment in the next room. (Recanati, 2007, pp. 149–150)

Similarly, someone might come to have a belief that she* desires $x$ based on her psychoanalyst convincing her thereof. In this case, the self-ascription of desire $x$ is not immune to error through misidentification; after all, the psychoanalyst might have gotten his clients somewhat mixed up and was in fact talking about someone else. However, in cases where someone self-ascribes desire $x$, such as the current desire for a cold beer, based on her current awareness of the desire in question, the self-ascription is immune to error through misidentification. So the distinction between circumstantial and absolute immunity seems misguided, insofar as whether a self-ascription is immune to error through misidentification always depends on the way in which it is formed. In cases where the way of gaining information underwriting the self-ascription in question is specific to the self, the resulting judgment will be immune. In cases where the relevant way of gaining information is not specific to the self, the resulting judgment will not be immune. Thus, as was already pointed out by Evans (1982), what matters is not the type of information (i.e., information regarding one's mental states versus information regarding one's bodily states) but the relevant ways of gaining information about the self.

What about the distinction between *de facto* and logical immunity? A judgment possessing logical immunity to error through misidentification is a judgment that is immune in all possible worlds. In contrast, a judgment that possesses *de facto* immunity is a judgment that is immune in the actual world but might not be immune in other possible worlds. Self-ascriptions of mental states, such as "I see a canary" or "I believe $x$," are thought to have logical immunity—provided that they are made based on what I earlier termed the internal mode of experience, that is, based on being introspectively aware of the mental state in question. This is because a mental state, such as a visual experience, a desire, or a belief, necessarily belongs to the subject experiencing the state in question, and there is no possible world in which this does not hold true. In other words, it is a question of conceptual necessity that an occurrent mental state belongs to the subject in whose mind it occurs (i.e., to the subject experiencing the state in question). Even if a subject was telepathically connected to the mental states of someone else, simply by virtue of the subject experiencing these mental states, they would become the subject's own. It is logically (or conceptually) impossible not to be the owner of a mental state you are experiencing; the awareness of being in a mental state is sufficient for making that mental state your own.[8]

In contrast, in the case of bodily states, it is often held that self-ascriptive judgments based on the experience of a bodily state—such as the state of one's legs being crossed—are merely *de facto* immune, because as a matter of fact, bodily experience is a way of gaining information that is specific to the subject of the experience in question; however, it is possible to think of a possible world in which this does not hold true. For example, if a subject was cross-wired to the body of another in such a way that the subject could not tell whether she was receiving information about her own body or that of another, the subject's self-ascriptions of bodily states, such as the state of having crossed legs, would not be immune to error through misidentification. It is logically possible to experience the bodily states of another, but it is not logically possible to experience the mental states of another.[9] Now, it might be thought that we need logical immunity to account for the possibility of self-consciousness. If so, contrary to what I have been arguing, bodily experience, which only gives rise to *de facto* immunity, would not suffice to ground self-consciousness.

Along those lines, O'Brien (2007) has argued that only agent's awareness, but not bodily awareness, can ground first-person judgments that are immune to error through misidentification in a strong sense, and hence only agent's awareness can ground our ability to self-refer in thought. Similar to my account, O'Brien also starts from the view that we should not attempt to account for the possibility of self-reference in terms of nonconceptual *self-representation*—though her reasons for this position differ from mine.[10] Indeed, consistent with my account, she also argues that what is crucial for an account of self-referential first-person thought is the *mode* of occurrence of certain experiences. In particular, she suggests that "the relevant mode of occurrence, which warrants the immediate self-ascription of the thought in question, should be understood in terms of agent's awareness" (O'Brien, 2007, p. 115). She further argues that "agent's awareness is the result of acting on the basis of an assessment of possibilities for acting" (p. 115). The basic thought here is that insofar as an agent is acting on the basis of assessing possibilities for action, these are necessarily possibilities *for the agent*. This is because such actions are under the agent's control; they are something the agent must *actively produce*. Accordingly, such actions immediately warrant self-ascription, without the need for observation, inference, or self-representation (cf. O'Brien, 2007, pp. 183–184).[11] Importantly, in O'Brien's view, agent's awareness is to be sharply distinguished from bodily awareness in terms of being able to ground self-consciousness. This is because we can imagine abnormal circumstances—that is, circumstances in which a subject is wired to, and receives information about, the body of another—in which bodily awareness can put a subject in a position to know that someone is F, without thereby knowing that she* is F. In contrast, it is impossible to know through agent's awareness the action of another as a producer. Thus there is a difference between bodily awareness and agent's awareness, such that in cross-wiring cases I could be quasi-bodily aware of the body of another—and hence of possibilities for movement of the other's body—but I could never have quasi-agent's awareness of another's action.[12] Importantly, in O'Brien's view, this fact is transparent to the subject; hence she speaks of transparent rather than logical immunity.

However, this proposal has a number of problems. First, it is not clear that we should seek the grounds for transparent (or logical, for that matter) immunity to error through misidentification in the distinction between a passive and an active phenomenology. O'Brien argues that agent's awareness is transparently immune to error because actions are something the agent must *actively produce*. However, I argued earlier that mental states such as perceptions, desires, or thoughts can also give rise to first-person judgments that are (logically) immune to error through misidentification. And these are not necessarily something the agent actively produces (at least not in the relevant sense). For example, desires are often experienced without being actively produced, and we are not usually in a position to actively produce or control our perceptions or emotions.[13] O'Brien might respond that her project was not to give a unified account of logical or transparent immunity; rather, her focus was on the distinction between agent's awareness and bodily awareness. Even so, the criterion of transparent awareness seems too strong. For one thing, as O'Brien (2012) herself points out, it is not clear that agent's awareness actually provides the grounds for transparent awareness. In O'Brien's (2007) view, any information received about the body of another by means that are phenomenologically similar to bodily awareness would be neutral between passive and active bodily movements, so that such information could at most ground existential knowledge claims about bodily movements, but not about action. After all, to know that someone's hand moved is not to know that someone moved her hand. Accordingly, if a subject came to doubt that she really moved her hand, she would thereby also lose her grounds for thinking that someone moved her hand (as opposed to the thought that someone's hand moved) (cf. O'Brien, 2012, p. 128). In other words, to know—"from the inside"—that someone moved her hand is necessarily to know that it is oneself who moved it. However, as O'Brien (2012) points out, good evidence indicates that a subject's perception is sensitive to the distinction between intentional action and mere bodily movement (see also chap. 6). If so, it stands to reason that bodily awareness might be sensitive to this distinction, as well. In that case, the kind of quasi-bodily awareness involved in cross-wiring cases could provide a subject with information that someone moved her hand while leaving her in doubt as to whether it was she*. Thus it is not clear after all whether we can sharply distinguish agent's awareness from bodily awareness in terms of being able to ground first-personal judgments that are transparently immune to error through misidentification.

In any case, and more importantly, why should we think that to explain the possibility of genuine "I"-thoughts, it must be logically or transparently impossible—as opposed to *de facto* impossible—to confuse an experiential source of knowledge about oneself with a source of knowledge about another? Indeed, the notion of transparent (or logical) immunity might be an altogether too demanding notion to be useful for a theory of self-consciousness. Recall that the aim is to give an account of the possibility of "I"-thoughts that does not fall into the problem of regress faced by subject–object models, on the one hand, and is psychologically real, on the other. As I have argued, in virtue of their mode of presentation, perception and bodily experience provide the subject with information that is necessarily self-related, such that in normal circumstances, if a subject is aware of property $F$ via perception or bodily awareness, she is warranted in self-ascribing property $F$. This warrant could only be called into question under extremely peculiar, counterfactual circumstances, such that the subject would be prevented from being able to take an apparently self-specific mode of experience at face value. But it is not obvious that to explain how subjects come to be able to form "I"-thoughts, we need to rule out such counterfactual circumstances.[14] After all, we have no reason to expect that a merely logically possible counterfactual scenario, such as the one in which a subject is cross-wired to the body of another, should be reflected in the cognitive architecture of organisms who have evolved in *this* world, that is, in circumstances where the possibility of such cross-wiring does not exist. Accordingly, we have no reason to think that such a scenario is relevant to the ability of organisms evolved in this world to form "I"-thoughts. Rather, it is plausible to assume that given that subjects are *de facto* entitled to form "I"-thoughts based on their perceptual or bodily experience, and given that such "I"-thoughts will be *de facto* immune to error through misidentification, this is all we need for an account of the possibility of self-consciousness in the sense of the ability to think "I"-thoughts.

Put differently, the reason that immunity to error through misidentification is an important marker for self-consciousness is that only if we allow the possibility of self-ascriptions that do not rely on identification, and hence are immune to error through misidentification, can we avoid the regress problem that besets traditional subject–object models of self-consciousness (see chap. 1). In other words, to explain the possibility of "I"-thoughts, we need an account of what it is to experience oneself "as subject," that is, an account that does not construe self-consciousness along the lines of object awareness. The view I am proposing here provides precisely such an account. According to this view, certain modes of awareness are self-specific due to the cognitive architecture of the experiential subjects in question. Accordingly, the self is not represented as an object in the representational content of these states; rather, the self-specificity is implied by the mode of presentation. But there is no reason to think that this self-specificity needs to hold in all possible worlds to explain the possibility of self-consciousness in *this* world. In a possible world in which the subject has a different cognitive architecture (e.g., a world in which its proprioceptive system is hooked up to several different bodies), the mode of presentation of bodily awareness would presumably not imply an essential self-relatedness. However, this does not show that the mode of presentation is not essentially self-related in this world (cf. Evans, 1982). If so, there is no reason to think that bodily awareness (and agent's awareness, for that matter) cannot be among the sources of self-consciousness, even if they are not logically or transparently immune to error through misidentification.

Indeed, in her 2012 paper, O'Brien relaxes her criterion of transparent immunity based on concerns that are similar to those I have just discussed. As she rightly points out, what matters for an account of self-consciousness is not so much logical (or transparent) immunity. Rather, what matters is the question of whether there are states with genuine first-person content, that is, states in which the subject is taken "as subject" rather than "as object." That said, O'Brien continues to maintain that in bodily awareness we take our bodies as objects, whereas in acting we do not take our bodies as objects. She holds that in the case of bodily awareness "there is an argument place in the structure of experience which is normally reflexive and fused to the experiencing subject, but which could, in peculiar circumstances, be left open by the subject" (O'Brien, 2012, p. 143), whereas in agent's awareness no such argument place exists. That is, she construes bodily experience along the lines of states with content (a) as described at the beginning of the chapter. However, if the arguments that I advanced here and in the previous chapter are right, the self is not represented in bodily experience; rather, it is implied in the mode. If so, bodily awareness is an instance of being aware of oneself "as subject," rather than "as object." In other words, there is no reflexive argument place, or no self-referring component, in the structure of bodily experience; rather, such experience is self-less—similar to perceptual experience. Accordingly, the content of perceptual as well as bodily experience should be construed along the lines of states with content (b). Bodily awareness would be an instance of being aware of oneself "as object" if it required self-identification. But according to the arguments I have presented, this is not the case.[15] This is not to say that we can never take our body as an object. Of course this is possible; we do this when we observe our bodies in the mirror, for example. However, bodily awareness is not the mode we employ when doing so. Put differently, O'Brien and I agree that self-consciousness should not be construed along the lines of the self-representationalist view. However, we disagree about whether bodily awareness is an instance of self-representation. O'Brien, like the self-representationalist, thinks that it is. Hence she also thinks, contra the self-representationalist, that it cannot ground self-consciousness. In contrast, I think that bodily awareness is *not* an instance of self-representation, and that precisely because of this, it *can* be among the sources of self-consciousness.

The conclusion we can draw from this is first that there does not seem to be any reason to think that, for the purposes of explaining self-consciousness, anything as strong as logical—or transparent—immunity to error through misidentification is required. Second, neither in agent's awareness nor in bodily awareness do we experience the body "as object"; rather, both are forms of being aware of oneself "as subject." Hence both are among the sources of self-consciousness.

None of this is to say that agency is not a crucial component of self-consciousness. Indeed, as we have seen, both the ascription and specification of representational content—including self-representational content—are closely tied to the fact that we are intentional agents. For instance, I have argued that in bodily experience a subject is aware of her body as a system of possible movements, and the content of perception is to be spelled out in terms of the affordances it presents to the subject. In that sense, the kind of experience that grounds "I"-thoughts is always agential. However, this does not seem to be the notion of agent's awareness that O'Brien has in mind when she contrast agent's awareness with bodily awareness. And the worry is that by drawing a sharp line between agent's awareness and other forms of awareness, such as bodily awareness, and by tying the notion of self-consciousness exclusively to the notion of agent's awareness, we end up drawing too narrow a boundary around the possible sources of self-consciousness. We are agents—and aware of ourselves as such—but we are also perceivers, bodily beings, and bearers of reactive attitudes.[16]

The larger moral we can draw from this discussion is perhaps that while the notion of immunity to error through misidentification is important for an explanation of self-consciousness, the distinction between different kinds of immunity might not be. In particular, the weight that some accounts put on the notion of logical (or transparent) immunity seems to be too strong. Indeed, as Coliva (2012) has argued, the notion of logical immunity might even lie at the heart of what she calls "illusions of transcendence about the self."[17] Coliva argues that only if we take seriously the distinction between de facto and logical immunity can we dispel such illusions—so in that sense, the distinction between logical and de facto immunity does seem to be important. However, on the flip side, this also seems to suggest that philosophers who fall prey to such illusions make precisely the mistake of taking the notion of logical immunity too seriously in their thinking about the self and self-consciousness. Accordingly, we should try to avoid this error.[18]

## 4.4 Conclusion

On the basis of this and the preceding chapter, we can conclude that the nonconceptual representational contents of perception and bodily experience do not represent the self. Accordingly, rather than trying to explain the possibility of self-consciousness by means of nonconceptual self-representational states, we should opt for a non-self-representational-ist theory that takes into account the mode of presentation of experien-tial states. States with nonconceptual content, such as those involved in perception and bodily awareness, can provide the basis for thoughts that are immune to error through misidentification. They can do so precisely because the self is not explicitly represented in the content of those states but rather implied by the mode. This fact can be made explicit through the application of the self-concept. Given that no identity judgment is involved in the transition from the implicitly self-related nonconceptual content of experience (where the self-relatedness is a function of the mode of experience) to the explicit representation of the self in (conceptual) thought (where the application of the self-concept makes explicit the role of the mode), the resulting self-ascription will be immune to error through misidentification. Thus, if we take into account the role of the mode of presentation, we can see that while so-called nonconceptual forms of self-consciousness are not in fact genuine forms of self-consciousness, because they do not fulfill the requirements of self-reference and immunity to error through misidentification, they can still be said to underwrite the "primi-tively compelling" judgments that give rise to genuine "I"-thoughts.

We can conclude that non-self-representationalist views are superior to self-representationalist views because they are able to avoid the problems encountered by the latter—in particular the reliance on the problematic subject–object model of self-consciousness and the problem of cognitive overload—and because they are able to provide us with an explanation of the phenomenon of immunity to error through misidentification. How-ever, these theories still fall short of a completely satisfying solution to the problem of self-consciousness. While they are able to show that the imme-diate self-ascription of mental and bodily states on the basis of consciously experiencing these states is warranted, they do not give an account of how a subject can acquire the self-concept required to perform such self-ascrip-tions. In other words, the theories discussed so far leave us with the ques-tion of what exactly explicit self-representation amounts to and how we are to make intelligible the transition from implicitly self-related information to explicit self-representation.

So my task for the remainder of this book will be to answer the ques-tion of how we are to understand the transition from implicitly self-related information to explicit self-representation. The next chapter spells out in more detail the distinction between implicit and explicit representation and presents a model of how one can get from the former to the latter. I argue that this transition occurs in a stepwise process of representational redescription, so that we can distinguish different levels of explicitness. Chapter 6 then turns to the question of how and when a conscious state containing implicitly self-related information turns into a truly self-conscious thought.

# 5 From Implicit Information to Explicit Representation

The conclusion of the previous chapter was that to provide a full account of the ability to think "I"-thoughts, we need an explanation of the transition from implicitly self-related information to explicit self-representation. To tackle this problem, it is necessary to spell out in more detail the difference between implicit and explicit representation—or, more accurately, the difference between implicit information and explicit representation— and to make intelligible the transition from one representational format to the other. Given that this transition in general—and the transition from implicitly self-related information to genuine self-consciousness in particular—can be assumed to proceed in degrees, explaining it will require an account of implicit and explicit representation that admits of different levels of explicit representation and thereby transcends the dichotomy associated with the distinction between nonconceptual and conceptual representations.

This brings us back to a point made in chapter 2. In the course of discussing arguments in favor of nonconceptual content, we saw that one argument supporting the existence of nonconceptual representations was based on the fact that perception—in contrast to belief—is situation dependent. That is to say, in contrast to conceptual representations, which allow for context-independent, domain-general, systematic, and productive thought, nonconceptual representations are context bound and do not allow for the generality and systematicity associated with conceptual representation. However, we also saw that when we turn to psychological arguments in favor of nonconceptual content, we need to recognize that there is considerable room in between simple stimulus–response behavior—which does not warrant the ascription of representational content at all—and full-fledged conceptual thought. In particular, various forms of intentional behavior and practical rationality call for the ascription of nonconceptual representational content, and the forms of representation associated with these vary with regard to their degree of context dependence or generality, respectively (cf. Hurley, 2006). Chapters 3 and 4 mainly considered relatively primitive forms of nonconceptual representation, such as basic perceptions and bodily experience. However, as became obvious in discussing the relevance of the notion of nonconceptual content for an explanation of intentional behavior and the practical means-end rationality that is associated with intentional behavior in chapter 2, there are likely to be more complex forms of nonconceptual representation than the ones discussed in the previous two chapters. These include forms of representation that are involved in different types of practical rationality, or in different types of social interaction (I discuss social interaction in detail in the next chapter). Presumably these different forms involve varying degrees of generality. Another way of putting this is by saying that we have good reasons to think that there are contents in between the categories of conceptual and nonconceptual. To account for this fact, we need a more fine-grained approach that is able to spell out the distinctions between different forms of representation and can make intelligible the transition from implicit information to explicit representation, and thus from context-bound and domain-specific to flexible and general forms of representation. Importantly, as I argued in chapter 2, nonconceptual content is to be spelled out in terms of the subject's possibilities for interacting with the world, that is, in terms of "knowledge-how," or procedural knowledge. Hence an account of the transition between nonconceptual and conceptual content must be able to make intelligible how content that is encoded in a procedural format, and guided by the norms of activity guidance and skill, can be transformed into content that is encoded in conceptual format and guided by the norms of truth.

It is the task of this chapter to present such an account. The account I present here draws on Dienes and Perner's (1999) "theory of implicit and explicit knowledge" and on Karmiloff-Smith's (1996) model of "representational redescription" (which, in turn, draws partly on ideas developed by Piaget).[1] The chapter begins by giving an account of the distinction between implicit and explicit representation in general terms, based on a natural-language understanding of the terms "implicit" and "explicit" (sec. 5.1). I then present the model of representational redescription to see whether and how we can make intelligible the transition from implicit to explicit representation. I also show how this model is supported by empirical observations (sec. 5.2). This leads to a more detailed consideration of the structure of different types of mental representation, the different levels of explicitness associated with them, and the relation between explicit representation and conscious access (sec. 5.3). Finally, I throw a first glance at the implications of the insights gained through this analysis for the relation between explicit representation, nonconceptual content, and self-consciousness (sec. 5.4). The next chapter then addresses the question of how the results established in this chapter allow us to better understand the nature and development of self-consciousness in particular.

## 5.1 Implicit versus Explicit Representation

In day-to-day language, we say that something is expressed explicitly when it is stated directly, such that the hearer or reader of a sentence can immediately grasp its meaning. According to Dienes and Perner (1999), for a given fact or state of affairs to be explicitly expressed in a sentence, the sentence needs to contain a verbal expression that directly refers to that fact or state of affairs. For instance, when I say, "It is raining in London," the location of the rain—namely, London—is represented explicitly. In contrast, something is expressed implicitly when it is conveyed but not stated directly, for instance, when the meaning has to be inferred from the context of what is being said. So if I simply say, "It is raining," I am still conveying that it is raining in a particular location (e.g., London), but the location is left implicit and must be inferred from the context, such as from my present location or from the conversation preceding the sentence in question.

If we transfer this ordinary-language analysis of explicit representation from linguistic to mental representations, we can say that a "fact is explicitly represented if there is an expression (mental or otherwise) whose meaning is just that fact; in other words, there is an internal state whose function is to indicate that fact" (Dienes & Perner, 1999, p. 736). So a fact or state of affairs is represented explicitly when the mental state in question contains a component (an expression) that directly stands for this fact or state of affairs. A related way of formulating this idea is to say that an explicit representation is a representation whose content is immediately graspable by the subject (Kirsh, 1990).[2] I discuss the relation between explicit representation and graspability (in the sense of conscious access) further in section 5.3.

In contrast, a fact or state of affairs is implicit in a mental representation when the mental state in question does not contain a component
that directly refers to the fact, but when the fact or state of affairs is still
conveyed as part of the contextual function of the mental state—that is,
when the fact is an "unarticulated constituent" of the representation. In
fact, we have encountered cases of precisely this implicitness due to contextual function in the previous chapters. For instance, the functional role
of bodily experience is to convey information about the bodily states of the
subject undergoing the experience. However, the essential self-relatedness
of the content of bodily experience is not explicitly represented; rather, it is
left implicit (i.e., unarticulated), as part of the mode or functional context
in which bodily experience operates (which is determined by facts about
the cognitive architecture of the subject). When I have a bodily experience of legs being crossed, it is necessarily my legs that I experience being
crossed. This is simply how bodily awareness works: it delivers information
about the experiencing subject's limbs and bodily states and nobody else's.
Importantly, this fact need not itself be explicitly represented precisely
because the function of bodily awareness is such that it necessarily delivers
information about one's own body. So the content of a bodily experience
need not contain an explicitly self-referring component for bodily experience to fulfill its functional role. Rather, the content of the bodily experience consists in the property alone, without containing a component that
refers to the subject role (cf. Recanati, 2007, 2012). Likewise, that visual
perception is necessarily subject relative is not part of the explicit content
of perception; rather, it is implicit in the sense that the agent-relative roles
of the objects that are being perceived are necessarily relative to the perceiving subject and not just any subject.

## 5.2   From Implicit Information to Explicit Representation:
## Representational Redescription

The question at this point is how we can get from implicit information
to explicit representation. This transition is, after all, what is required for
the emergence of the ability to entertain genuine "I"-thoughts. To answer
this question, it is helpful to consider the development of human cognitive abilities. After all, as I argued in chapter 2, one of the motivations for
introducing the notion of nonconceptual content was that infants, even

though they clearly represent the world, lack conceptual abilities. So if we want to know how it is possible to get from the self-related information implicit in nonconceptual representations, such as those involved in perception and bodily awareness, to the concept of a self, it might be instructive to consider how children master the transition from nonconceptual representations to the kind of conceptual representations that we find in mature adult humans.

According to the model of "representational redescription" proposed by the developmental psychologist Karmiloff-Smith, an infant starts with certain innate domain-specific predispositions, where a domain is "the set of representations sustaining a specific area of knowledge: language, number, physics, and so forth" (Karmiloff-Smith, 1996, p. 6). These broad domains can also contain microdomains (e.g., the language domain contains the microdomain of pronoun acquisition, the mathematical domain contains the microdomain of counting, and so on). According to Karmiloff-Smith, the information implicit in these domains and microdomains is turned into explicit representation through a reiterative process by which "information already present in the organism's independently functioning, special-purpose representations, is made progressively available, via redescriptive processes, to other parts of the cognitive system" (p. 18).[3]

The model posits at least four levels of representation. At the first level (I), information is encoded in procedural form. Information in this form remains implicit and is unavailable to other operations in the cognitive system. In a subsequent reiterative process of representational redescription, the information is transformed into increasingly abstract and less-specialized but more cognitively flexible formats of representation. These latter representations can then be used for cognitive operations that require explicit knowledge. At the second level (E1), information is available as data to the system, in the form of "theories-in-action," although not necessarily to conscious access and verbal report. At the third level (E2), information becomes available to conscious access, but not yet to verbal report. And finally, at the fourth level (E3), information is re-represented into a cross-system code, which allows it to become verbally expressed (Karmiloff-Smith, 1996, p. 23). Thus the process of representational redescription involves the recoding of information that is stored in one representational format into another, such that each redescription is a more condensed version of the previous level. As a result, the information becomes increasingly explicit and can be used increasingly flexibly. The advantage of explicit representations is that they allow for more cognitive flexibility and domain generality. In contrast, implicit or procedural representations enable fluid and automatic interactions with the environment but remain domain specific. Notice that this model implies that the representational system is much more complex than the dichotomy between nonconceptual and conceptual content would suggest (p. 22).

So the theory of representational redescription attempts to account for the transition from domain-specific (i.e., nonconceptual) to increasingly domain-general (i.e., conceptual) representations. As an example of domain specificity—we will encounter additional examples in what follows—recall the example of the chimpanzee Sheba discussed in chapter 2. Sheba is able to represent numerosity and is able to map the numerosity of a stimulus to numeric symbols (at least up to the number 4). Moreover, she is able to use her ability to judge the numerosity of a stimulus to her advantage in a specific context or domain, that is, in a task where she is confronted with symbols; call this domain A. Nonetheless she is unable to use the same information about numerosity in a different but relevantly similar context or domain, that is, when confronted with candy; call this domain B. Although the information she uses in domain A is relevant to her success in domain B, and although she is able to reliably use this information in domain A, she seems unable to use it in domain B. That is to say, she possesses representations that enable her to employ skills of practical reasoning to achieve a specific goal—"choose the symbol that indicates smaller numerosity to obtain the reward with the bigger numerosity"—but these representations are restricted to domain A (cf. Hurley, 2006). In contrast, mature humans would find it easy to switch between the two domains and use the information available successfully in both domains, as they are able to represent information in a format that allows them to generalize between the two tasks.[4] According to the model of representational redescription, this requires a recoding of the information that is implicit in the procedures that are activated when confronted with a task in a certain domain into a format that allows this information to be cross-transferred into a different domain. Let us now take a closer look at how the process of representational redescription is thought to occur according to Karmiloff-Smith's theory.

## 5.2.1   Level (I): Implicit Representation

As mentioned before, at the first, implicit level (I), information is encoded in procedural form—in procedures for processing and reacting to environmental input—with the information embedded in the procedures remaining implicit, that is, unavailable to other operations within the cognitive system. Reconsider the example from chapter 2 regarding infants' abilities to parse a visual scene by analyzing three-dimensional surface arrangements and following the motion of perceived objects. As we have seen, studies have shown that four basic principles underlie infants' perception of visual scenes: boundedness, cohesion, rigidity, and "no action at distance." Yet for infants to show the described expectancy responses, it is not necessary that they explicitly represent these principles. Rather, the four principles are presumably "encoded in the form of procedures for responding to environmental stimuli" (Karmiloff-Smith, 1996, p. 77). This means that the fact that the visual system of infants operates on the basis of these principles can be inferred from the behavior—the expectancy responses—shown by the infants. Nonetheless the explicit content of the infants' representations need not contain objects and their properties, such as "there are two objects and they cannot pass through each other," nor do they need to contain an explicitly self-referring component, for reference to the self is implicit in the mode of perception. In other words, the information is *in* the system, but it is not available *to* the system as such. This is consistent with the analysis presented in chapter 2, according to which the nonconceptual content of perception is to be spelled out in terms of the organism's possibilities for interacting with the environment, or in terms of knowledge-how.

Consider the example of a skilled football player. The representations that guide the player's movements are such that they allow for fluid and fast play, but they are inaccessible to other parts of the cognitive system, for instance, the language system, and so cannot be expressed verbally. A particular procedure as a whole may be available as data to the cognitive system, but its component parts are not (cf. Karmiloff-Smith, 1996, p. 20). For instance, the procedure of passing a ball as a whole will be accessible to the player, so that when she is prompted to make a pass, she is able to do so immediately. However, the individual steps involved in the procedure are not explicitly represented and hence will be inaccessible to the player, so that she is not in a position to explain each individual step involved in the process of passing. Of course, in certain circumstances, such as when a person switches from the role of player to the role of trainer, she might be able to explain the individual steps (to some extent); but for this to work, the information would have to be available in a format different from the one that is at play when the footballer is fluidly executing movements. In the role of trainer, the subject will access explicit and conceptual representations, while in the role of player, the information remains implicit.

### 5.2.2   From Implicit Information to Explicit Representation: Level E1 Representations

So at the level of implicit representation, information is encoded in procedural form. The content of representations at this level is restricted to special-purpose mechanisms, which is to say, it is not generalizable. Accordingly, it falls under the class of nonconceptual content. However, as indicated earlier, given the range of different types of intentional behavior, ranging from relatively simple perceptual discriminations to fairly demanding practical and theoretical inferences, we should distinguish between different degrees of specificity or generality. To make sense of certain types of intentional behavior, it is necessary to assume the existence of mental states with representational content that is more explicit, and thus more general, than the content involved in procedural representations. Nonetheless such content does not yet need to reach the level of generality associated with conceptual content. According to Karmiloff-Smith's model, such states would have to be located at the first level of explicit representation (E1).

Examples of such level E1 representations are so-called "theories-in-action." The following example (which is discussed in Karmiloff-Smith, 1996, chap. 3) will help to elucidate the transformation of implicit information to level E1 explicit representations, and to lend empirical support to the theory.[5] This example involves children's ability for block balancing and demonstrates the children's passage from behavioral mastery (level I), to "theories-in-action" (level E1), and finally to consciously accessible and verbally expressible theories about gravity (level E2/E3).

In the study, a group of four- to nine-year-olds were asked to balance a series of different blocks on a thin metal support (Karmiloff-Smith, 1984). The blocks were such that some had their weight evenly distributed, and hence balanced at their geometric center, while others were filled with lead weights at one end so that their balance was off center. A third type of block also balanced off center but had a visible weight glued to one end. The results showed that four- and five-year-olds easily solve the task of balancing the different blocks, as do eight- and nine-year-olds. Six- and seven-year-olds, however, perform poorly in this task (such U-shaped performance trajectories are commonly found in developmental psychology). How is this—at first glance rather surprising—result to be explained?

According to Karmiloff-Smith, four- and five-year-olds simply pick up each block and move it along the support, using proprioceptive feedback to find the right point of balance. They do not make any specific selection among the blocks; that is, even if they just succeeded in balancing a block of a certain type, they will not choose a block with a similar appearance to apply the information that they have just gathered in the next trial. Instead they apparently treat the balancing of each block as a separate problem. As Karmiloff-Smith points out, this age group seems to strive for behavioral mastery rather than engaging in theorizing, and their behavior is underwritten by representations that contain implicit (i.e., procedural) information about the relation between weight distribution and point of balance contained in the positive and negative proprioceptive feedback.

In contrast, the study found that six- and seven-year-olds place every block at its geometric center and therefore only manage to balance blocks with evenly distributed weight. This suggests that their behavior is no longer purely input driven, as was the behavior of the younger group. Rather, it seems to be mediated by a "theory-in-action," which causes these older children to ignore the proprioceptive feedback they receive. Why should this be so? One possibility is that once children reach behavioral mastery, they begin to direct their focus toward their own internal representations of the world, which enables them to extract general features from their representations. In other words, they begin to engage in theorizing. For example, in the block-balancing task, based on their previous experience with objects, six- and seven-year-olds might have extracted a common feature that holds for most, if not all, objects they have encountered: namely, that objects balance symmetrically along their length at their geometric center. According to Karmiloff-Smith, this is the core of the reduced, re-represented information that was previously implicit in the procedural knowledge that children acquire by interacting with the environment (Karmiloff-Smith, 1996, p. 86). In other words, the children generate a proto-theory based on previously stored independent representations and begin to use this proto-theory to shape incoming data. Thus, once children have reached a certain level of behavioral mastery, they begin to form a "geometric-center theory," albeit one that they cannot yet conceptualize or express verbally. This can be inferred on the basis that the children's behavior conforms to their "geometric-center theory," even though they are not yet able to verbally state the content of this proto-theory, or to explain why they place the blocks in the way they do. Nor should we assume that this kind of theorizing occurs consciously.

According to Karmiloff-Smith, once a proto-theory is in place, it takes a while for children to give up on it again. Accordingly, they ignore data that contradict the theoretical commitments they have established—such as the proprioceptive feedback in the task at hand. This is because, "like the behavioral mastery at level I, the consolidation of a theory at level E1 takes time developmentally. The theory itself must be consolidated before counterexamples are explained via a different theory" (Karmiloff-Smith, 1996, p. 86).

### 5.2.3 From "Theories-in-Action" to Full Explicitness: Levels E2 and E3

The problem of counterexamples is solved only when children reach a new level of explicitness and begin to develop the correct theory. According to Karmiloff-Smith, for this particular case, this occurs at eight to nine years, when children acquire explicit knowledge about geometric centers and begin to develop a naive but correct theory. From now on, children again succeed in balancing uneven blocks, replicating the behavior of the youngest group. However, the representations underlying their behavior are very different from those at play in the performance of the youngest group. At this age, children reach the next level of representation (E2), which provides them with conscious access to the theoretical commitments that were previously implicit in their behavior, and ultimately enables them to adjust these commitments based on new data, as well as to verbally state these commitments and explain and justify their behavior (level E3).[6] It is only at this level that information is recoded into a fully domain-general code, allowing for a transfer of knowledge to other domains and thus for a flexible and creative application of this knowledge. So it is only at this level that we can speak of fully conceptualized knowledge, whereas at level E1, one would still speak of nonconceptual, or perhaps partially conceptualized, knowledge.

It is important to note that in Karmiloff-Smith's view, implicit and explicit representations can exist alongside each other in different modalities. So while children might possess explicit representations about the balancing of objects at a certain age—alongside their implicit representations that guide their intuitive interaction with these objects—when it comes to other modalities, they might not yet have made that transition. Moreover, implicit representations are not simply replaced by explicit representations. While explicit representations require a re-representation and redescription of the original implicit representations, this does not mean that the implicit representations are abolished during the process of redescription. For example, while a football player who wants to become a trainer or write a book about how to play football needs to find a way to access the information that is implicit in her skillful play so that it can be expressed verbally, this does not mean that she thereby loses her ability to play. Both the implicit representations that guide her fluid and automatic movements and the explicit representations that enable her to write books and teach others remain present, albeit in very different formats. Similarly, when you ask the six-year-olds from the block-balancing task to close their eyes, they no longer have difficulties balancing the different blocks. This demonstrates that they still have access to the proprioceptive feedback that guides the behavior of four- and five-year-olds.

Likewise, humans who acquire an explicit self-concept do not thereby lose the agent-relative information that is implicit in their perceptual and action-guiding systems. Nor does one lose the implicit (or level E1 explicit) representations that underlie one's fundamental social skills in the process of developing an explicit theory of mind, as I discuss in the next chapter.[7]

While Karmiloff-Smith's model provides a way of conceptualizing the different types of nonconceptual content that underwrite different types of practical rationality, so far, her proposal leaves open what mechanisms drive the developmental process through which implicit information is translated into explicit forms of representation, or why some organisms progress to more explicit representations (and ultimately conceptualization) while others do not. Although I take this to be largely a matter for future empirical research, I suggested in the previous section that the ability for metacognition plays an important role in this process. According to Karmiloff-Smith, representational redescription involves a process of directing attention away from external information and toward one's own internal representations. Accordingly, representational redescription seems to involve the ability to represent one's internal states for the purpose of extracting information, so as to redescribe it into a more abstract format; in other words, it involves metacognition. If this is right, only organisms capable of metacognition would also be capable of representational redescription. Accordingly, animals that display intentional behavior that suggests the presence of explicit (E1 or E2) representational content should also perform well in experiments that test metacognitive abilities, and vice versa. Notice that metacognition in the sense required here does not imply conscious processing, explicit metarepresentation, or self-identification. In other words, the child need not be aware of being in a state that is directed at a lower-level internal state for the child's cognitive system to extract information from the lower-level state. Rather, metacognition in this context can itself be thought of as an "implicit, procedural form of reflexivity" (Proust, 2006, p. 263). After all, metacognition as it is being understood here is meant to be a necessary prerequisite for explicit metarepresentation (cf. Proust, 2006), so metacognition cannot be identical with metarepresentation. (See chap. 7 for a further discussion of metacognition.)

Alternatively, Mandler (2004) has proposed that an innate mechanism, which she calls "perceptual meaning analysis," in conjunction with the innate characteristics of our perceptual systems, can account for the kind of theorizing behavior just described, or, as she calls it, for "conceptual learning" (Mandler, 2004, p. 67). The notion of perceptual meaning analysis is in some ways related to the notion of representational redescription, albeit with important differences. In particular, while Karmiloff-Smith seems to suggest that the process of representational redescription relies on dynamics that are internal to the system and based on the direction of attention away from external information toward internal representations once a child reaches behavioral mastery, Mandler posits an active analysis of incoming, external data, prior to and independent of behavioral mastery (Mandler, 2004, p. 75). In Mandler's view, the mechanism of perceptual meaning analysis is thought to result in the creation of image schemas, which are taken to be analog representations that summarize spatial relations and movements in space (p. 79). It is important to note that in her view, image schemas are not simplified pictures but reflections of the structure of spatial information that can be derived from vision, audition, and touch, as well as from movements created by the child, based on the restrictions placed on the child's sensory system and the relevance of movement.

They are analog, but also relatively abstract, which allows them to be used in generalizations from familiar to new contexts.

However, it is not my aim here to adjudicate between these two theories. In fact, it seems plausible that both the striving for behavioral mastery and subsequent theorizing based on metacognition, as well as an active analysis of perceptual input prior to, and independent of, behavioral mastery, may play a role in the transition from implicit information to explicit representation. Moreover, it is not my aim to delve deeper into a discussion of possible mechanisms involved in representational redescription at this point.[8] Rather, what matters for our purposes is that there are good reasons—both at a theoretical and at an empirical level—to assume the existence of different levels of explicitness, which suggests that our philosophical analysis should go beyond the simple dichotomy suggested by the distinction between nonconceptual and conceptual representations. Indeed, as I argue in the following section, the distinction between implicit and explicit representation does not map neatly onto the distinction between conceptual and nonconceptual representation, though there is a relation between these distinctions. Further, the model presented here implies that implicit and explicit representations can exist alongside each other, playing different roles for the explanation of behavior. I return to this point in chapter 6.

Notice also that I am not claiming that the levels described by Karmiloff-Smith are the only way to characterize different levels of explicitness, nor that they are applicable to all domains in exactly the same manner. It is entirely possible that, depending on the cognitive domain under investigation, fewer or more levels can be distinguished. In fact, in the next chapter, I propose my own framework for distinguishing between different levels of self- and other-representation, which differs to some extent from the levels described by Karmiloff-Smith. Nonetheless what makes Karmiloff-Smith's theory interesting for our purposes is that it provides us both with reasons to think that we can and indeed should distinguish between different levels of explicit representation, and with a model of how to view the transition between these different levels. Independent of whether all the specific details of her particular theory will prevail, applying her general framework to my discussion here provides a substantial theoretical advance, which enables us to move beyond the simplified dichotomy of conceptual versus nonconceptual representations and thus to create a more precise and accurate view of human and nonhuman cognition.

## 5.3   Explicit Representation, Conscious Access, and (Non)Conceptual Content

What is the relation between explicit representation, conscious access, and conceptual content? The model of representational redescription postulates that a certain level of explicitness is required for conscious access to the information in question. So for someone to consciously access a given fact or state of affairs, one must explicitly represent it. Indeed, it is no coincidence that the distinction between implicit and explicit representation is often associated with the distinction between unconscious and conscious processing (e.g., Schacter, 1987). In fact, many researchers assume that we can simply take implicit representation to mean unconscious representation and regard explicit representation as a stand-in for conscious representation. Similarly, we saw in chapter 2 that nonconceptual content is noncompositional, or structure implicit. For instance, a nonconceptual visual experience of a tree represents certain features of the tree—in terms of the subject's abilities for interacting with the tree—without explicitly representing the individual object (e.g., "tree"), its property (e.g., "having green leaves"), and the predication relation between them (e.g., "this tree has green leaves") as such. An explicit representation of these components requires concepts. A subject can only explicitly represent that a property, for instance, "having green leaves," can be predicated of an object, for instance, "a tree," if the subject has the concept of a tree and of the color green (and if the subject is able to represent a predication relation).[9]

So a relation clearly exists between explicit representation, conscious access, and concepts. However, the arguments in chapter 2, as well as the analysis from the previous sections, also showed that simple dichotomies, such as the dichotomy between unconscious and conscious processing, or the dichotomy between nonconceptual and conceptual representation, do not do justice to the complexity of the human representational system. For example, the four-year-old children in the block-balancing task presumably have a conscious perceptual experience of the blocks they are trying to balance and of the bodily feedback involved in solving the task. Nonetheless they rely on procedural, implicit knowledge rather than on explicit representations to master the task. Explicit representations enter only at a later stage, when the information implicit in the stored representations of their perceptual and proprioceptive experiences is redescribed into a different format—that is, into the format of a "theory-in-action" of the point of balance of objects, and ultimately into an explicit, conceptualized theory. Likewise, as we have seen in the previous two chapters, I can have a conscious, and even conceptual, visual perception that is such that certain aspects of the perceptual content—for instance, its self-relatedness—remain implicit in the representation. Similarly, at the level of E1 representations, there is a certain level of explicitness—in the form of a "theory-in-action"—but the theoretical commitments that seem to constitute this "theory-in-action" are not such that they can be consciously accessed or verbally expressed. These facts cannot adequately be captured just by appealing to the difference between conscious and unconscious processing, or between conceptual and nonconceptual content.

This is because any given conscious experience has different constituents, each of which can either be implicit or explicitly represented. For instance, according to the standard analysis of propositional attitudes (e.g., "I know that there is a tree in front of me"), we can distinguish between the content or proposition, standardly expressed by means of a "that" clause (in the example: "that there is a tree in front of me"); the propositional attitude, for example, knowledge versus belief versus desire, and so on (in the example: "know"); and the holder of that attitude (in the example: "I").[10] Accordingly, as Dienes and Perner (1999) point out, there are three main types of explicit representation that are determined by which of the three constituents of a propositional attitude is represented explicitly. We can distinguish between (1) explicit content but implicit attitude and implicit holder of the attitude; (2) explicit content and attitude but implicit holder of the attitude; and (3) explicit content, attitude, and holder of the attitude. According to Dienes and Perner, we often entertain propositional attitudes with explicit content, while the attitude remains implicit in the functional role of the representation. For example, a desire has a different functional role in my cognitive architecture than a belief, a fear, or a wish, even when I don't explicitly represent this fact, that is, even when I'm not aware of this fact. However, the propositional attitude—the functional role of the representation—itself can also be explicitly represented, as in when I metarepresent the fact that I am currently entertaining a belief as opposed to a wish, imagination, or worry. This metarepresentation can still leave the holder of the attitude implicit, though. It is only when I self-ascribe the representation by forming an "I"-thought that the holder of the attitude (i.e., myself) becomes explicitly represented.

For example, say that at this moment I see a red lamp on my desk. To consciously access the content of this visual experience (e.g., to draw or report on the lamp), I must explicitly represent the red lamp. However, I need not explicitly represent the fact that this is a visual perception, or that it is I who has this experience, to explicitly represent the lamp. Rather, as we saw in the previous chapters, the subject of perception can remain an "unarticulated constituent" of the representation. However, for me to be aware that I am entertaining a visual representation, the information that was previously implicit in the functional role that my representation played—that is, in the mode of presentation—must be redescribed into an explicit format. And the same holds for the fact that it is I (rather than someone else) who is seeing the lamp. This fact is implicit in the fact that I do indeed have the experience, but it must be made explicit for me to consciously access it, such that I can then form an "I"-thought, as in "I see a red lamp."

Similarly, there is no reason to think that the children in the block-balancing task are explicitly representing their propositional attitude or themselves as the holder of this attitude to start forming a proto-theory. Rather, they only need to represent the content of their representations of objects to extract common features (such as the feature that objects tend to balance at their center).

This picture is further complicated because the content of a propositional attitude itself also admits of different levels of explicitness. According to Dienes and Perner, we can distinguish between four main components of content, each of which may be either explicitly represented or left implicit. For instance, in the case of an individual seeing a tree, we can distinguish between (1) properties, such as "F," "being a tree"; (2) individuals, such as "b," "particular individual in front of me"; (3) the predication of the property to the individual, such as "Fb," "this is a tree"; (4a) the temporal context, such as "I see a tree in front of me now"; and (4b) factuality versus fiction, such as "it is a fact of this world at time t, Fb," "it is a fact that this is currently a tree." Accordingly, for any mental representation, we can ask of each of these components whether they are implicit or explicitly represented.

This shows—in line with the arguments brought forward in chapters 2 and 3—that although nonconceptual representations are structure-implicit, and fully structure-explicit representations require concept possession, the distinction between nonconceptual and conceptual representations as such (similar to the distinction between unconscious and conscious representations) is not fine grained enough to do justice to these different levels or degrees of explicitness. In other words, there must be contents that fall somewhere in between the categories of nonconceptual and conceptual, that is, contents that are what we might call partially conceptual.

## 5.4   Summary and Conclusion: Explicit Representation, (Non)Conceptual Content, and Self-Consciousness

So what does all of this mean with regard to our discussion of so-called nonconceptual forms of self-consciousness? As we have seen in the previous chapters, simply pointing out that nonconceptual forms of representation like ecological perception and bodily experience contain self-related information does not suffice to show that they constitute forms of self-consciousness. This is because perception and bodily experience contain implicitly self-related information rather than constituting explicit self-representation. We can now use the framework of different levels of explicit representation provided by Dienes and Perner (1999) and Karmiloff-Smith (1996) to give a more detailed account of the distinction between implicit information and explicit representation and the transitions between them. What we ought to take into account, according to this framework, is that mental representations consist of different components, and we can ask of each component whether it is explicitly represented or whether it remains implicit.

In the case of a conceptual representation in the form of a reflective, verbally expressed "I"-thought, such as "I believe that there is a red lamp in front of me," not only are the components of the perceptual content, namely, the object "lamp" and its property "red" explicitly represented, but also the propositional attitude "to believe" and the subject of the attitude "I" (which explicitly refers to the utterer of the statement). Accordingly, this is an example of a maximally explicit representation. In contrast, in the case of an experiential state with perceptual content that can be specified in terms of nonconceptual content (e.g., the perception of a visual scene), we have a representation that is explicit in the sense that there is conscious awareness of the visual scene in question, but the content of the perception itself is structure implicit in the sense that there is no explicit representation

of object, predicate, or predication relation, let alone of propositional attitude or of subject of perception. The latter two remain unarticulated; that
is, they remain implicit in the mode or functional role of the presentation.
This state can thus be said to possess a certain level of explicitness despite
having nonconceptual content. In between are states with varying degrees
of explicitness, including states where the content is structure explicit (e.g.,
where an object is represented as an object with specific predicates), but the
attitude and the holder of the attitude are left implicit; or states in which
content and attitude are explicitly represented (i.e., metarepresentations),
but where the holder of the attitude is still left implicit. (We will encounter
examples of this in the next chapter.)

According to Karmiloff-Smith, for information that is implicit in the cognitive system to become explicitly represented, and thus generally accessible to the system, it must be recoded into a different format. With regard to
self-consciousness, this means that the implicitly self-related information
contained in perception, bodily experience, and intentional action needs
to be reformatted such that it becomes available to the subject for conscious access and, ultimately, for the conceptualization of this information
*as being about* the subject. In this manner, the subject learns to think of
herself as a perceiver and (embodied) agent and to ascribe various bodily
and mental states to herself. This requires not only an explicit representation of the states that are being self-ascribed but also the development of a
self-concept.

The crucial question at this point is how this process of representational
redescription of implicitly self-related information into explicit self-representation is to be understood, and how the subject acquires a self-concept.
I turn now to this question.

# 6 Self and Others, or The Emergence of Self-Consciousness

The task of this chapter is to apply the insights gained so far to the problem of self-consciousness. The arguments presented in the previous chapters lead to two conclusions: First, I argued that world-directed action and perception do not require explicit self-representation. This raises the question of when explicit self-representation does become necessary. In other words, it raises the question of when we move from consciousness to self-consciousness. In the following, I argue that this need arises only in the context of intersubjectivity, and therefore a theory of self-consciousness must take into account the relation between the awareness of oneself and the awareness of others. Second, I argued that the transition from implicitly self-related (i.e., self-concerning) representations to explicit self-representation requires a process of representational redescription, and we can distinguish different levels of explicitness within this process. Based on these two insights, this chapter presents a model according to which implicitly self-related information is turned into explicit self-representation via a process of an increasingly complex differentiation between self and other.

The term "intersubjectivity" will be used in a rather loose sense and taken to indicate any kind of social cognitive process that involves two or more subjects, independently of whether these subjects have an explicit understanding of each other *as* subjects. There are good philosophical arguments for restricting the term to contexts that involve a mutual recognition of each other as intentional subjects, and some philosophers might even want to restrict it to contexts of mutual recognition as moral agents. Nonetheless I adopt the usage that is most common in both the psychological and the contemporary philosophy of mind literatures. I take this strategy to be unproblematic, because it will always become clear in discussing various phenomena throughout the chapter whether we are dealing with a situation of mutual recognition or with more primitive forms of social cognition.[1] The chapter begins with a brief historical prelude (sec. 6.1). Although it falls outside the scope of this book to provide a detailed account of the relevant historical positions (indeed, doing so would require writing a book of its own), and although the terms "self-consciousness" and "intersubjectivity" are not always used in the same way by different authors, I take it to be helpful to remind ourselves of the parallels between contemporary debates and historical treatments of the subject. This will enable us both to gain an appreciation of the multidimensionality of the phenomenon of intersubjectivity and to see how some of the historical ideas presented in this overview resurface in current discussions—albeit not usually by means of explicit reference. After this brief overview, I present some theoretical arguments to explain in further detail why there is what I take to be a constitutive relation between self-consciousness and inter-subjectivity (sec. 6.2). The theoretical discussion is followed by a detailed description of different levels of implicit and explicit representation of the mental and bodily states of self and other, supported by empirical obser-vations (sec. 6.3). This leads to a discussion of different theories of social cognition (i.e., simulation theory, theory-theory, interaction theory, and the narrative practice hypothesis) in which I suggest that, rather than see-ing these accounts as competing, we should see them as complementing one another, with different theoretical approaches needed to account for the different levels of self–other representation described in the preceding section (sec. 6.4). Finally, I consider additional evidence from the study of autism to further support the existence of a strong relation between self-consciousness and intersubjectivity (sec. 6.5), before presenting a summary and some concluding remarks (sec. 6.6).

## 6.1  Historical Prelude

The idea that an important, even constitutive, relation links self-conscious-ness and intersubjectivity is by no means new. In fact, historically, the idea of linking self-consciousness and intersubjectivity can be interpreted as an alternative—or rather a necessary complement—to the view that self-con-sciousness can only be explained by referring to some kind of prereflective self-consciousness. As seen in chapter 1, according to the subject–object model, to become conscious of itself, the subject has to reflect or represent itself to itself. However, this model turned out to be circular. Consequently some authors argued that the ability for self-reflection necessarily presupposes the possession of a more fundamental form of self-consciousness, that is, a prereflective form of self-consciousness. This is, as we have seen, the view defended by proponents of the so-called Heidelberg School, as well as by some phenomenologists, and, in analytical philosophy, by proponents of theories of nonconceptual self-consciousness. However, as I argued in the previous chapters, if this view is to be plausible, (a) it must be interpreted not in terms of nonconceptual self-representation but in terms of the non-self-representationalist view; and (b) it must be complemented with an account of how the ability to think explicit "I"-thoughts can emerge out of states with nonconceptual, implicitly self-related content. One way of going about this is by arguing that the subject comes to think explicit "I"-thoughts—and hence becomes self-conscious—through the encounter and interaction with other subjects. Arguably, this was the route taken by thinkers such as Fichte, Hegel, Sartre, Mead, and later Habermas (to name just a few).[2]

In the tradition of German idealism, Fichte can be seen as the first to propose such a view.[3] He argued—in line with thinkers of the Heidelberg School—for the need for a prereflective self-consciousness (Frank, 1991a).[4] However, Fichte also held that self-consciousness becomes reflective and intentional only through intersubjective encounters. The basis for his argument is his concept of a "summons" (*Aufforderung*). In a nutshell, according to Fichte, the self-conscious subject essentially thinks of herself as a free agent who acts in light of her reasons for action.[5] So free agency is to be understood in terms of reasons responsiveness. In Fichte's view, the experience of oneself as a free agent—that is, as an agent who is acting for reasons—is not possible without one's mind instantiating the concept of a "summons," that is, of being addressed. This concept, in turn, cannot be thought without being thought of as originating in a rational being that is distinct from oneself. In other words, I become self-conscious insofar as I am conscious of another's consciousness as addressing (summoning) me, thereby providing me with a *reason* for a free action, as opposed to a mere *cause* of my actions (Wood, 2006). The element of a summons is also retained in Hegel's work, though for Hegel, intersubjectivity is characterized by a mutual struggle for recognition during the course of which each subject becomes conscious of itself (as famously expressed in his slave–master allegory).

On the part of phenomenological approaches to the topic, we again find positions that speak in favor of a prereflective self-consciousness, such as the view defended by Sartre (1943). On the other hand, in Sartre's phenomenology, reflective self-consciousness is the result of intersubjectivity, which, for him, is characterized by a constant mutual objectification and, consequently, self-alienation (as famously explicated in his description of the other's gaze in *Being and Nothingness*).[6] Take his famous example of someone spying through a keyhole. It is only when the spy hears the footsteps of another approaching that the spy becomes (immediately, and guiltily) aware of what she is doing. Before this moment, she is immersed in her activity without any explicit awareness of herself as an agent. But as the spy is about to feel the other's gaze on her, she becomes "fixed" in this gaze and thereby self-conscious. In other words, according to this view, I acquire reflective self-awareness in considering how I am being perceived—and turned into an intentional object—by the other. However, according to other thinkers in the phenomenological tradition, most notably Merleau-Ponty (1962), self-reflection of this kind presupposes an awareness of one's visibility to the other. This awareness, in turn, is thought to be based on the prereflective, proprioceptive-kinesthetic sense of my body, and a common "corporeal schema" between self and other that enables my recognition of the similarities between my own and the other's body (Gallagher & Zahavi, 2010). Thus Merleau-Ponty seeks to locate the relation between self-consciousness and intersubjectivity already at the prereflective, bodily level, rather than at the more abstract level of mutual recognition and summoning.[7]

Yet other thinkers consider linguistic communication as being at the heart of the way in which we understand ourselves and others.[8] For instance, according to Mead (1934), linguistic communication involves an encounter between mutually acknowledging selves. This leads to the constitution and recognition of the self as an "I" and a "me." "The 'I' is the response of the organism to the attitudes of others; the 'me' is the organized set of attitudes of others which one himself assumes" (Mead, 1934, p. 175).[9]

It seems, however, as though intersubjective interaction in the form of linguistic communication already presupposes the existence of mutually recognizing subjects that can access shared contexts. In other words, to be capable of entering into communicative interactions in the first place, certain forms of social cognition must already be in place, and these seem to remain unaccounted for by philosophical theories that focus exclusively on linguistic forms of social interaction. For instance, to understand gestures or linguistic symbols as referring to a certain object, the subject must already be in a position to understand the communicative intent of the other (Bogdan, 2009; Grice, 1957). Thus the focus on linguistic forms of intersubjectivity seems to be too narrow to do full justice to the phenomenon of social cognition. Moreover, if the relation between self-consciousness and intersubjectivity were to be located at the linguistic level alone, this relation could not help solve the problem of self-consciousness. As shown in chapter 1, whether or not they explicitly take into account the intersubjective dimension of language, theories of self-consciousness that seek an explanation at the linguistic level, such as the one proposed by Tugendhat (1979), remain reliant on elements that they cannot account for. Indeed, as we will see shortly, there are forms of prelinguistic intersubjectivity on which linguistically mediated forms of intersubjectivity can build.

I will end my brief excursus into the history of theories of intersubjectivity at this point, for it is not within the scope of the book to provide a detailed interpretation of historical philosophical positions dealing with the relation between self-consciousness and intersubjectivity. The points to take away are simply that the insight that such a relation exists has a philosophical history, and that self-consciousness and intersubjectivity are multifaceted phenomena, which should be reflected in a proper analysis of the relation between self-consciousness and intersubjectivity. In particular, we need an analysis that can explain the progression from more primitive, bodily, and nonlinguistic forms of intersubjectivity, which were the focus of attention for phenomenologists like Merleau-Ponty, to the kinds of linguistically mediated phenomena that build the focus of thinkers such as Mead or Tugendhat. Fortunately, given the progress in philosophical theories studying these phenomena, as well as in developmental psychology and other cognitive sciences dealing with this area, we are now in a reasonable position to pursue this project and to provide an analysis of different intersubjective phenomena that not only accounts for different levels of intersubjectivity and self-awareness but also pays due respect to and draws on the insights gained from the empirical sciences. This project is my focus in the following sections.

## 6.2 Arguments for the Constitutive Relation between Self-Consciousness and Intersubjectivity

Let me begin by explaining in more detail what I take the relation between self-consciousness and intersubjectivity to be. The core idea is that self-consciousness is essentially a contrastive notion. Just as explicit reference to a place or time only makes sense for a being that is aware of the existence of other places or times, so explicit reference to the self only makes sense for a being that is aware of the existence of other subjects, to whom it intends to draw a contrast.

As we have seen in the previous chapters, perception and bodily experience contain self-related information, that is, information that in fact relates to the subject. But for this to hold, the subject itself does not need to be explicitly represented. Perception and bodily experience are action guiding because the information received through their mode of presentation is necessarily self-related. This is why, as discussed in chapters 3 and 4, theories of nonconceptual self-consciousness are correct in arguing that the question of who the subject of perception or bodily experience is cannot arise with respect to these representational states. There simply cannot be a question as to who is seeing the book on the table when I have a visual perception of the book on the table (and so it is pointless to even ask the question), just as there can be no question as to whose legs are crossed when I have a bodily experience of legs being crossed. Crucially, as we have seen, for this to hold, the self does not need to be a component of the explicit representational content of perception and bodily awareness. Rather, it can be conceived of as an "unarticulated constituent" of perception. Now, a version of the principle of parsimony requires that we should not ascribe complex representational states to an organism when simpler states can just as well account for the behavior in question. Accordingly, since explicit self-representation is not required for perception and bodily experience to fulfill their functional roles, we should not describe perception and bodily experience as being self-representational states. As long as an organism is only in the business of interacting with objects in the world by representing their location in relation to itself, their locations relative to each other, and the affordances provided by them, the organism does not need an explicit self-representation, and hence, for reasons of cognitive economy, we should not ascribe self-representational content to the organism in question.

So when is it necessary for us to represent ourselves explicitly at all? Evidently, this becomes necessary only when we begin to recognize and interact with other subjects. Once a cognitive organism recognizes that there are other cognitive organisms in the world, whose behavior depends not just on their physical properties but also on their representational states, new strategies for predicting the behavior of these other organisms come into play, strategies that—in contrast to the kinds of interaction afforded by material objects—require the ability to recognize and ascribe representational states to other beings (Beckermann, 2003). This will in turn enable the organism to contrast its own representational states with those of others. For instance, it may come to recognize that another organism can or cannot see what it itself can see, and act accordingly (e.g., by hiding a valuable piece of food). In doing so, the organism is contrasting its own perceptual states with those of another, thereby gaining awareness of its own states as being its own (cf. Pauen, 2000). Moreover, once an organism recognizes the existence of other cognitive systems with mental states of their own, it may also come to recognize that sometimes it can become the intentional object of these states, and it can use this information to take into account the way in which it appears to others, or what others know about it, and adjust its behavior accordingly. If so, the organism will begin to develop an explicit self-representation as a being with specific bodily as well as mental properties that can be represented by others (cf. Beckermann, 2003; Rochat, 2009).

In addition to this argument, which is based on considerations from cognitive economy, the case for the constitutive relation between self-awareness and awareness of others can also be made based on considerations about what it means to possess the self-concept and to be able to ascribe mental and bodily states to oneself in the form of "I"-thoughts.

The ability to think "I"-thoughts requires the ability to ascribe mental and bodily predicates to oneself by means of applying the first-person concept. As argued in the previous chapters, concept possession is required to think genuine "I"-thoughts, for it is by applying the self-concept in self-ascribing one's mental and bodily states that one makes explicit the essential self-relatedness that is implicit in the mode of perception and bodily or agentive awareness. Notice that concept possession is not an all-or-nothing phenomenon. Rather, as concept acquisition proceeds in degrees, via a process of representational redescription, there can be a partial grasp of concepts. Now, the ability to ascribe mental and bodily predicates to one-self is intimately tied up with intersubjectivity. As Strawson (1959) and later Evans (1982) have pointed out, the ability to ascribe predicates to oneself requires the ability to also ascribe these predicates to others (see also Campbell, 1994). According to Strawson, our mental-state concepts are such that they have first- and third-person application. Consequently, a grasp of these concepts implies the ability to ascribe them to others as well as to oneself.[10] A similar argument can be made based on Evans's Generality Constraint (see chap. 2). According to this constraint, a subject cannot be credited with the possession of a particular concept unless she is able to apply the concept in a variety of different instances. So a subject cannot be credited with possessing a given mental or bodily state concept unless she is able to apply the concept both to herself and to others who are not herself.[11] Note that it would not be sufficient for a subject to attribute predicates to herself at different times (*pace* Bermúdez). This is because a subject could hardly be credited with grasp of, say, a concept such as pain if she could only recognize pain states in herself but never in others (or vice versa).[12]

What this means is that insofar as the ability to explicitly refer to oneself in thought is ultimately a conceptual ability—albeit one that can be acquired in degrees—to give an account of the transition from implicitly self-related thoughts to explicit self-representation (and thus of the transition from mere consciousness to self-consciousness), we need to explain how a subject comes to acquire the ability to recognize herself as a subject among others and to ascribe mental and bodily states both to herself and to others.

Bermúdez (1998) argues that the ability to ascribe psychological states to oneself and others can be traced back to basic, and nonconceptual, social cognitive abilities. While I agree with him in principle, I disagree with the details of his proposal.[13] In particular, he claims that social cognitive abilities such as the ability for joint visual attention, joint engagement, and the ability for social referencing, which begin to emerge at around nine months, indicate the ability to ascribe mental states to another in a nonconceptual way. The problem with this proposal is that its interpretation of these early social cognitive abilities is much too rich. Indeed, and against his own commitments, Bermúdez describes these basic forms of interaction in terms that seem to require representational abilities that are conceptual, rather than nonconceptual. Consider the case of joint visual attention, which occurs when the infant attends to an object based on her perception of the gaze of another, or, conversely, attempts to direct the other's gaze to an object of interest. Bermúdez claims that the following representational contents need to be attributed to the infant to explain the ability to engage in this kind of interaction: "(1) The infant recognizes, 'Mother wants me to look where she is looking,'" or alternatively "(2) The infant recognizes, 'Mother is trying to get me to look where she is looking'" (Bermúdez, 1998, p. 259). However, it is hard to see why the ability to form thoughts with this kind of complexity should not count as a conceptual ability. Moreover, as we will see in the next section, the ascription of such complex representational abilities does not adequately describe what is going on in these basic forms of social interaction. Rather, much simpler explanations for the behavior in question are available, which imply neither the ability to explicitly distinguish self from other nor the ability to ascribe mental states (see sec. 6.3.2).[14] Thus Bermúdez's account, in addition to the problem of being self-representationalist, which was criticized in chapters 3 and 4, also suffers from a lack of differentiation with regard to the different levels or degrees of explicitness involved in the development of social cognitive abilities.[15]

In the following, I provide an account of the gradual transition from implicitly self-related to explicit self- and other-representation. Based in part on the analysis from the previous chapter, my argument here distinguishes between different levels of implicit and explicit self- and other-representation. Further, I show how the distinction between these levels is supported by empirical evidence from the cognitive sciences (in particular developmental psychology). This will provide us with a model of the relation between self-consciousness and intersubjectivity and its development that is more precise and differentiated than existing accounts.

It is important to be clear that although I take the symmetry thesis to be correct (i.e., the thesis that a constitutive relation exists between the ability to think about oneself and the ability to think about others), this does not imply that the way in which a subject self-ascribes mental and bodily states (where the "I" is used "as subject") is always the same as the way in which a subject ascribes the same states to others, as is claimed by Carruthers (2011). This is because, as discussed in both chapter 1 and chapter 4, the self-concept is not to be understood in terms of self-identifying information, as the self is not presented to itself in experience. Rather, possession of the self-concept implies sensitivity to the role of the mode of presentation involved in conscious experience, which guarantees the essential self-relatedness of the content of experience. This role is made explicit when the subject forms an "I"-thought by applying the self-concept. This is what explains why, in these cases, the resulting "I"-thought is immune to error through misidentification. In contrast, in the case of other-ascriptions, the subject does rely on identifying information.

However, this does not mean that self-ascriptions are infallible or even particularly authoritative, nor does it mean that we cannot also make self-ascriptions based on self-observation and self-identification. It might even be the case that we often self-ascribe mental states based on the observation of our behavior (for a detailed discussion of empirical evidence that suggests that this might be the case, and that this procedure often leads to mistakes, see Carruthers, 2011). The claim is just that such self-ascriptions—where the "I" is used "as object"—cannot be the only form of self-consciousness or self-knowledge (see chap. 1). In particular, awareness of oneself qua perceiver, embodied being, or deliberating agent in its paradigmatic form is not awareness of oneself "as object" and does not rely on self-observation or inference.

Let me also be clear that I do not claim that self-consciousness always requires a social context. Rather, intersubjectivity is required for the *acquisition* of the ability to think about oneself; once this ability is in place, a subject can, of course, form self-conscious thoughts in the absence of any kind of social interaction.

The account defended here is compatible with, and partly draws on, the Intentional Relations Theory proposed by Barresi and Moore (1996), whose distinctions between different types of intentional relations I adopt here. As we saw in the previous chapter, mental representations can be analyzed as involving three main constituents: the subject being in a mental state with representational content, the content of the representational state, and the intentional relation between subject and representational content. Among intentional relations, we can distinguish three main types. First, there are action- or goal-oriented intentional relations, for instance, the intention to grasp an object. Second, there are emotional intentional relations, such as fear of a dangerous situation. Third, there are epistemic intentional relations, such as having a belief, perceiving, believing, or knowing. Naturally, these often come in various combinations, for instance, when I believe that there is a deep cliff in front of me, and this belief is associated with my fear of cliffs, or when I feel content upon having reached my goal of grasping a desired object. Moreover, it is not clear whether all intentional attitudes fit into this simplified schema (it is not so clear how to categorize desires, for instance). Nonetheless it makes sense to draw these distinctions because not only do there seem to be conceptual differences between them (difficulties in the differentiation of individual cases notwithstanding) but, as we will see in a moment, empirical evidence also indicates the existence of different systems underlying the ability to represent and share different types of intentional relations.

## 6.3   Levels of Self- and Other-Representation

To even begin to recognize that other beings have bodily and mental states like oneself, and to contrast these with one's own (or to take into account the mental states of others that are about oneself), a being first needs to be in a position to recognize the similarity between itself and others. This requires that the being is able to match the information it receives about others, from the third-person perspective,[16] with the information it receives about itself, from the first-person perspective (Barresi & Moore, 1996; see also Tomasello, 1999). This is by no means trivial. After all, I have devoted considerable space throughout the book to discussing the differences between the access to one's own states "from the inside," compared to information about the outside world or about other subjects (see esp. chaps. 1, 3, and 4). In the case of others, I need to rely on the observation of their movements, their facial expressions, or the direction of their gaze and their position relative to external objects to know about their intentions, their emotions, or their perceptions. In contrast, to know about my own intentional states I only need to attend to the intentional objects of my representational states (cf. Barresi & Moore, 1996). For example, I need to observe another's gaze direction and spatial relation to an external object (e.g., a cup) to know whether the other can see it. In contrast, to know that I can see the cup, I just need to direct my attention to the cup. This difference is what Tugendhat's (1979) notion of "epistemic asymmetry" refers to. It is also the basis of what is sometimes called the "transparency" of self-knowledge. Given this asymmetry, how can we nonetheless realize that there are other beings just like ourselves?

A first requirement for this realization is that other subjects must be recognized as intentional agents; that is, one must adopt the "intentional stance" (Dennett, 1987) toward them. Empirical results show that in humans, a number of—presumably innate—mechanisms seem to predispose infants to recognize members of their own species and to adopt the intentional stance toward others. For instance, already at birth, infants show a preference for attending to facelike objects (Johnson & Morton, 1991), as well as for attending to human speech (Mehler, Jusczyk, & Nilofar, 1988). Moreover, young infants are already able to distinguish and respond differently to animate and inanimate objects (Legerstee, 1992; Opfer & Gelman, 2010).

However, the preferential treatment of social stimuli in comparison to stimuli from objects in the environment on its own does not yet imply an understanding that these social stimuli originate in beings that are similar to oneself. For this understanding to be possible, there needs to be some kind of matching mechanism that enables subjects to match input from self and other—that is, to match first- and third-person information—and to integrate the information in such a way that the resulting representation can be applied to both self and other. Otherwise, first- and third-person information would always be treated separately, and no common representational and, ultimately, conceptual scheme could develop (Barresi & Moore, 1996; see also Strawson, 1959).

Some authors have suggested that the neural basis for this matching mechanism lies in the mirror neuron system (Barresi & Moore, 2008). Mirror neurons were first detected in the premotor area F5 of macaque monkeys (Rizzolatti et al., 1996) and fire both when an action (such as grasping for an object) is perceived and when that same action is executed, thus bridging the "gap" between agent and perceiver (Gallese, 2001; Gallese, Keysers, & Rizzolatti, 2004). Evidence from functional imaging studies points to a similar system in the human brain (Iacoboni et al., 1999; Keysers & Gazzola, 2009).

### 6.3.1  Primary Intersubjectivity or Self–Other Matching

Empirical evidence further suggests that in humans some such matching mechanism is in place from a very early age and might even be innate, such that "infants, even newborns, are capable of apprehending the equivalence between body transformations they see and the ones they feel themselves perform" (Meltzoff, 1990, p. 160).[17] For instance, five-month-old infants have been shown to be able to detect the relation between their leg movement and a real-time video display of that movement, presumably based on their detection of contingencies between the visual and proprioceptive information (Bahrick & Watson, 1985). Bahrick and Watson showed that infants preferentially fixated on a noncontingent display of the leg movements of a peer or of themselves, compared to a display of movement that was perfectly contingent with their own leg motion. Moreover, as Meltzoff and Moore (1977) have demonstrated, newborns seem to be able to imitate the facial gestures of adults, such as tongue protrusion and mouth opening, shortly after birth. This seems to suggest that infants are able to match visual information about the facial expression of others with the proprioceptive information they receive about their own faces.[18] Accordingly, some researchers have argued that the ability to match first-person and third-person inputs is innate, though the innateness of this matching mechanism remains controversial.[19] Nonetheless it is uncontroversial that some such mechanism exists, explaining humans' general abilities for imitation and providing the necessary basis for the development of a common conceptual scheme for self- and other-attributions of intentional relations.

Importantly, this is not to say that at these very early stages of social interaction, infants already have an explicit representation of a self–other matching. The automatic matching mechanisms, which enable the detection of multimodal contingencies and seem to underlie infants' ability for imitation, require neither a conscious mapping, nor a distinction between self and other, nor a representation of the other (or of oneself) as an intentional agent. In fact, it seems plausible that at the very early stages of self–other matching indicated by neonate imitation, there is no differentiation between self and other whatsoever. Cases of early infant imitation are thus comparable to the phenomenon of emotional contagion (which I discuss further in sec. 6.3.3), where perceiving an emotional expression by another person causes the experience of the same emotion in oneself, without evidence of a differentiation between the other's emotion and one's own.[20]

### 6.3.2 Secondary Intersubjectivity

It is often pointed out that social interactions and the representations associated with them reach a new quality at around nine to twelve months. As was observed earlier, at this age, infants enter into contexts of shared attention and intentionality. Within the developmental literature, these are also called forms of "secondary intersubjectivity," in contrast to the so-called forms of "primary intersubjectivity" described in the previous section (Trevarthen, 1979). At this stage, infants begin to coordinate their object-directed behavior with their person-directed behavior; that is, they move from dyadic forms of interaction to triadic forms of interaction. For instance, they will follow the gaze or the pointing gesture of another person to an object of mutual interest, thus exhibiting "shared attention" toward that object (Tomasello et al., 2005). This suggests an implicit (i.e., procedural) understanding of others as perceivers, and the existence of a mechanism that integrates the first-person information about one's own perceptual situation with the third-person information received about another's perceptual situation.

Moreover, infants at this age begin to develop social referencing, which is to say that they use emotional information from their caregivers to regulate their own behavior in situations that are perceived as threatening (e.g., Feinman, 1982). One example of such social referencing (which is also discussed in Bermúdez, 1998) is the visual cliff paradigm used by Klinnert et al. (1983). A visual cliff is a table made of glass and visually divided into two halves. On one half, a pattern is placed immediately below the glass plate so that this half appears opaque, while on the other half a similar pattern is placed at a distance from the glass plate, so that there appears to be sudden drop-off. When children reach the drop-off point, they will stop and look to their caregivers. If the caregiver looks happy, the child will continue its crawl, but if the caregiver looks worried, the child will refuse to move forward. This suggests a matching between first- and third-person emotional intentional relations, such that children adopt the emotional state they perceive their caregivers to be expressing to regulate their own behavior.

Children at this age also begin to display communicative gestures, such as proto-imperative and proto-declarative pointing gestures (Bates, 1979). Proto-imperatives function as a form of nonverbal request to a partner, for instance, to obtain an object that is out of reach. In contrast, proto-declaratives are more akin to nonverbal comments on a situation, for instance, when the pointing gesture is used to inform another about the location of an object. While the proto-imperative is an attempt at influencing what another person does and is thus directed at the other's action-related intentional relations, the proto-declarative is directed at the other's attention or other mental states and is thus directed at the other's epistemic intentional relations (Barresi & Moore, 1996; Karmiloff-Smith, 1996, chap. 5). So while the proto-imperative seems to suggest an implicit understanding of others as agents, the proto-declarative seems to suggest an implicit understanding of others as perceivers and believers. Interestingly, while nonhuman (but human-raised) primates have been found to use proto-imperatives to some degree, they do not seem to use proto-declaratives (Tomasello, 2008). This suggests that while chimpanzees have an implicit understanding of others as agents and can use this understanding instrumentally to their advantage, they have either no understanding or no interest (or neither) to influence the mental states of others outside instrumental contexts. That is, in contrast to humans, chimpanzees do not seem to engage in the sharing of information or cooperation outside instrumental contexts.[21]

However, *pace* Bermúdez (1998), neither shared attention, nor social referencing, nor the ability to use proto-imperatives and proto-declaratives requires the explicit attribution of mental states to others. In other words, there is no indication that children at this age understand mental representations *as such*; that they are able to understand, for instance, that mental states can misrepresent. Rather, the representations of others as agents, perceivers, and bearers of emotions that are involved in these contexts of secondary intersubjectivity are implicit in the procedures involved in different types of social interaction (cf. Karmiloff-Smith, 1996, chap. 5). They seem to be a form of "knowing-how" (e.g., knowing how to direct someone's attention to an object), rather than a form of "knowing-that" (e.g., knowing that the other is a perceiver, intentional agent, etc.). Nor do these abilities require an explicit differentiation between self and other. During episodes of shared attention, there is a matching of first- and third-person information, in the sense that infants perceive the gaze orientation of the other while simultaneously experiencing proprioceptive feedback from their own head or eye movements and seeing the object of shared attention (Barresi & Moore, 1996). Likewise, in the case of social referencing, the infant sees another's emotional expression and adopts a corresponding emotional attitude. So shared intentionality enables the integration of third-person information about another's appearance, expression or behavior with first-person information about one's own intentional relations. Nonetheless infants understand the intentional relations associated with these types of social interaction only to the extent that they actually engage in episodes of shared intentionality, and this engagement does not require an explicit differentiation between first- and third-personal sources of information (cf. Barresi & Moore, 1996).

Thus the understanding of others as intentional agents in these cases seems to be located at the implicit, procedural level. This analysis is further supported by the fact that these representations seem to be domain specific; as we saw in the previous chapter, for information that is implicit in domain-specific behavioral procedures to become available to other parts of the cognitive system, it needs to be redescribed into a more explicit format.

At ten months of age, when infants are in the process of developing understanding of communicative actions such as pointing, and of states of social attention such as mutual gaze, these developments are not closely related: a child may master one of these domains while making little progress in the other. … Moreover, ten-month-old infants reliably follow a person's gaze to the object at which she is looking and look at an object to which she is reaching, but they fail to connect these two abilities so as to predict that a person will reach for the object to which she looks. (Spelke, 2009, pp. 168–169; see also A. T. Phillips, Wellman, & Spelke, 2002)

This suggests that infants at this age fail to integrate their implicit understanding of others as agents with their implicit understanding of others as perceivers who share their own experiences of the world (cf. Spelke, 2009, p. 169). The representations that are implicit in different social interactions must first be transformed into a more general, explicit format before infants can develop an integrated understanding of others—and consequently of themselves—as agents, perceivers, *and* bearers of emotions.

### 6.3.3   Mirror Self-Recognition, Self-Conscious Emotions, and Empathy

For a subject to realize that other subjects are distinct beings with their own mental states, she needs to be in a position not only to match third-person information from others with first-person information from her own experience in current episodes of shared intentionality, but also to distinguish between her own and the other's intentional attitudes. Moreover, she needs to understand that she can become the intentional object of another's intentional attitude.

A first step toward an explicit distinction between self and other might be the ability for selective imitation, which begins at around fourteen to eighteen months. In contrast to the neonate imitation discussed earlier, the kinds of imitation displayed at this age are clearly intentional. First, children distinguish actions that are being marked as intentional from those that are being marked as unintentional (e.g., by an exclamation of "Oops!") and imitate only the intentional actions (Carpenter, Akhtar, & Tomasello, 1998). Second, in the case of the imitation of unsuccessful actions, children will display the intended action rather than the unsuccessful one (Meltzoff, 1995). Third, children choose the most rational means when imitating. For instance, they will imitate an unusual action, such as pressing a light button with one's head, only when the experimenter's hands are free—indicating that he intentionally chose his head—but not when his hands are tied (Gergely, Bekkering, & Király, 2002). This suggests that children are able to grasp the other's intention, rather than simply matching a movement. Moreover, it also seems to suggest the ability to distinguish between the affordances of the other and their own. After all, they do not simply copy the other's movements but rather select a way to imitate the other's intended action based on the possibilities for movement at their disposal, even when these differ from the possibilities for movement at the disposal of the other (e.g., by using their hands rather than their head in cases where the hands of the experimenter were tied). Yet the ability for selective imitation does not yet suggest an explicit awareness that one can become the object of another's intentional attitude—for instance, that one can be perceived by another.

A possible indicator for such an awareness is the ability for mirror self-recognition. This ability is standardly tested with the help of the rouge test (or mirror test), in which the subject is marked with a red spot on her face and then observed in front of a mirror (Amsterdam, 1972; Gallup, 1979). If the subject attempts to touch or remove the spot on her face, this is taken to demonstrate that she recognizes herself in the mirror—which might imply an understanding that one can be perceived by others, that is, from the third-person perspective (Barresi & Moore, 1996). This ability typically emerges in children at around eighteen to twenty-four months. However, notice that the ability to pass the mirror test in and of itself does not yet demonstrate an awareness that one can be perceived by others (i.e., an awareness of oneself as a subject that can become the intentional object of another's mental state). Rather, this ability can also be explained on the basis of visual-kinesthetic matching (Mitchell, 1993, 2002), or the ability to use "novel, displaced visual feedback on [one's] physical state and behavior" (Heyes, 1994, p. 916). However, contrary to what is often claimed, this does not yet imply a concept of the self or an ability to ascribe mental states to oneself and others.

That said, mirror self-recognition is generally found to be accompanied by the development of so-called secondary or self-conscious emotions, such as embarrassment and coyness (Lewis et al., 1989).[22] This may suggest that at this age the child indeed begins to have a sense of being a potential object of perception and that the child's emotional attitudes are affected by this understanding, thus demonstrating an integration of her first-person experience with an understanding of the third-person information that others have about her (which she can henceforth also have about herself by gaining information about herself in the ways that others do, such as by observing herself in a mirror). In other words, this may imply that the child now begins to understand that she is a subject that can be observed by others, just as she is in turn able to observe the behavior of others (cf. Barresi & Moore, 1996). To use Rochat's (2009) terminology, it therefore seems plausible that from this point onward, the child has "others in mind." If so, at this point the child begins to fully appreciate herself as a subject among other subjects (i.e., other beings with intentional attitudes), and thus we can speak of intersubjectivity in a more substantial sense of the term.

This interpretation is consistent with findings that children at this age also begin to display a general understanding that someone else's attitude or orientation toward an object can differ from their own. For instance, eighteen-month-old children understand that others may like something (e.g., food) that they themselves do not like, and vice versa (Repacholi & Gopnik, 1997). In addition, they begin to understand that they may see something that someone else does not see, and vice versa; that is, they begin to engage in so-called level 1 perspective taking (Moll & Tomasello, 2006).

Signs for an understanding of the emotional intentional relations of others as being different from one's own also begin to emerge during the second half of the second year. One example is the development of empathy (Hobson, 2002). Younger children will typically get distressed themselves and seek comfort when they perceive expressions of distress by others, thus exhibiting signs of emotional contagion. In contrast, children during the second half of the second year—while still showing some signs of distress themselves, indicating that there is some first-personal experience of the relevant emotional state—will try to do something to comfort the other, thus demonstrating an understanding that the emotion "belongs" to the other.[23] Much later, once children develop an explicit theory of mind, they will even be able to show sympathy, which is the ability to understand and be sensitive toward the mental states of others without experiencing them from the first-person perspective.[24]

Taken together, these findings suggest that from about eighteen months onward, the child begins to differentiate between self and other and to attribute intentional relations differentially to self and other. Before this age, the child has *de facto* access to the mental states of others and engages with them, but she need not explicitly represent these states *as* belonging to others, for this access occurs during episodes of shared intentionality, in which the child herself also experiences the intentional relations she shares with her partner (Barresi & Moore, 1996). So for shared intentionality to occur, it is sufficient that intentional relations are in fact shared; the child does not have to explicitly represent her own mental states as being distinct from those of others. In contrast, in the case of empathy, although the child will share some of the emotion of the other (as evidenced by the child's behavioral signs of distress), the comforting behavior that she directs toward others shows that she represents the emotion as belonging to the other. Likewise, the child learns to represent perceptual states and preferences as belonging either to herself or to others. Thus, during the second half of the second year, the child begins to represent the fact that others have mental states as well and that these may differ from her own. However, these representations do not yet have to be in a conceptual or linguistic format. Rather, they seem to be in what Karmiloff-Smith has termed E1 format, which is to say that they might constitute something like a "theory-of-mind-in-action,"[25] without that theory being conceptualized or consciously accessible in terms of explicit propositional attitude ascriptions as such.

### 6.3.4 Explicit Mentalizing and Theory of Mind

Although mirror self-recognition, self-conscious emotions, perspective taking, and empathy indicate an understanding of being a subject among other subjects (i.e., bearers of mental states), none of the abilities described so far yet implies the existence of an explicit theory of mind. To truly appreciate mental states as such, a subject must be able to distinguish intentional attitudes from intentional content and must understand that mental states can misrepresent.

This ability is demonstrated when children pass so-called false-belief tasks, usually around four years of age. It is only at this stage that children reach the ability to explicitly represent belief states as such, that is, to represent them both as states that can be held by others and differ from one's own beliefs, and as states that can misrepresent and thus differ from reality. In other words, it is only at this stage that children acquire a conceptual understanding of mental states. In a classical false-belief task designed by Baron-Cohen, Leslie, and Frith (1985), based on a modification of the original task designed by Wimmer and Perner (1983), a child watches a scene involving a puppet, Sally, and another puppet, Anne. Sally hides an object (e.g., a marble) in a basket while Anne is watching. When Sally temporarily leaves the room, Anne transfers the marble to another container (e.g., a box). The child who is watching the scene is then asked where the marble really is, and where Sally will look for it when she returns. To pass the test, the child has to distinguish between what he himself knows about the marble's location and Sally's (false) belief about the location of the marble. So the child has to be able to distinguish his own propositional attitude regarding the location of the marble from Sally's propositional attitude regarding the location of the marble. He has to be able to think, for example, "I know that the marble is in the box" versus "Sally thinks that the ball is in the basket." That is to say, the ability to pass the false-belief test requires an explicit representation of propositional content, propositional attitude, and holder of the attitude (i.e., me versus Sally).

The child also has to know that Sally's behavior will be determined by her internal mental state, rather than by reality (cf. Karmiloff-Smith, 1996, chap. 5). This implies that the child must now be able to integrate his knowledge about Sally as an agent with his knowledge about her as a perceiver and believer, for Sally's belief state will be determined by where she saw the ball being hidden earlier, and this in turn will determine how she acts. As we saw in section 6.3.2, this kind of integrated representation is lacking at the level of implicit representations of intentional relations. At the implicit level, representations of others as perceivers, agents, and bearers of emotions seem to be domain specific and cannot yet be transferred to, or integrated with, information from other domains. But once the information has been recoded into an explicit format, it can be generalized and applied across domains, thus leading to a more complete understanding of other persons—and of oneself.

Recently, the widely held consensus that children do not possess an understanding of beliefs as propositional attitudes that can misrepresent before the age of four years has been challenged by experiments that seem to demonstrate a false-belief understanding in children as young as eighteen months (Buttelmann, Carpenter, & Tomasello, 2009), fifteen months (Onishi & Baillargeon, 2005) or even younger (Surian, Caldi, & Sperber, 2007; Kovács, Téglás, & Endress, 2010). If so, this would also pose a problem for the view defended here. For example, using a "violation-of-expectation" paradigm, Onishi and Baillargeon's study showed that when confronted with nonverbal false-belief tasks, infants' looking behavior indicates the correct location (i.e., where the other person falsely believes the object in question to be) long before they are able to correctly answer questions about that location. However, it is unclear how to interpret these results. Some authors (e.g., Perner & Ruffman, 2005) have argued that the results reported by Onishi and Baillargeon can be explained in terms of relatively simple behavioral rule following rather than in terms of an understanding of false beliefs *per se*. Moreover, an explicit understanding of mental states implies a number of abilities, such as an appreciation of the intentionality of beliefs, their complex relations to other mental states as well as to behavior, the possibility of misrepresentation, and the ability to attribute propositional attitudes differentially to different holders of such attitudes. The findings reported in the studies I just cited do not demonstrate the ability to reason about mental states in this sense (cf. Apperly & Butterfill, 2009; Butterfill & Apperly, 2013; Spaulding, 2011). For example, the findings in question do not imply an understanding that beliefs can be caused by and in turn influence other mental states. That is to say, these results do not imply that infants at this age possess the concept of belief (though they might indicate the emergence of a partial grasp of the concept). Rather, these results might instead hint at a nonconceptual understanding of certain aspects of behavior that can function as proxies for beliefs (e.g., "an agent will search for an object where she has last encountered it"), as has been suggested by Apperly and Butterfill (2009; see also Butterfill & Apperly, 2013).[26] Alternatively, they might be based on the ability to implicitly adopt and track another's visual perspective, without necessarily distinguishing this perspective from one's own (which might account for the results reported in Surian et al. 2007 and Kovács et al., 2010, in particular). So while the behavior displayed in these experiments might go beyond a mere behavior reading, it does not yet amount to a full grasp of mental-state concepts (cf. Apperly & Butterfill, 2009; Butterfill & Apperly, 2013; Spaulding, 2011). This interpretation is compatible with the kinds of abilities described in the previous section and would also explain why infants at this age still fail on explicit, verbal false-belief tasks. In contrast to "violation-of-expectation" paradigms, verbal false-belief tasks require an explicit, conceptual representation of the other person's belief—in contrast to one's own—in a format that allows for conscious access and verbal report.[27]

It is noteworthy that at around the same time they pass verbal false-belief tasks, children also begin to display a number of related cognitive abilities (as described in Rakoczy, 2008). For instance, they begin to solve unexpected-content tasks (Perner, Leekam, & Wimmer, 1987). In this task, children are presented with a box (e.g., a candy box) and asked what they believe is inside. They are then shown the real contents of the box, for example, a pen. Afterward they are asked what another child will think is in the box, and what they previously thought was in the box. Children younger than four years often fail on this task, but by the time they reach four years, they will generally give the correct response. Moreover, children at this age also begin to distinguish appearance from reality (Flavell, Flavell, & Green, 1983). That is, they begin to distinguish what an object seems to be (e.g., a sponge that looks like a stone) from what it really is (i.e., a sponge). They also begin to understand visual perspectives, that is, they begin to display level 2 perspective taking; for instance, they are able to tell whether a drawing looks upside-down to an observer who is sitting opposite them (Flavell et al., 1981). Finally, they master tasks involving intentional deception, for example, by deceiving a "nasty" puppet, with whom they (or a "friendly" puppet) are competing for a reward, either through deceptive pointing or by telling a lie (Sodian, 1991, 1994). What all these tasks have in common is that they imply an understanding of epistemic perspectives as being different from reality and from one's own perspective, and of the fact that it is the content of subjective mental states provided by the respective perspective that is guiding the actions of individuals. In other words, the child learns to explicitly ascribe propositional attitudes to others and to use these as premises in predicting and explaining the behavior of others (Perner, 1991; Rakoczy, 2008).

Thus, at this level, children possess a theory of mind that is explicitly represented in conceptual and linguistic format. This is further supported by the strong connection between linguistic abilities and the understanding of beliefs and folk psychology (for an overview, see de Villiers, 2005; Zlatev, 2008). For example, deaf children who are not exposed to sign language from very early on show a delayed understanding of the (false) beliefs of others compared with hearing children and children who have signing parents (Peterson & Siegal, 1995). Moreover, longitudinal studies indicate that language development predicts theory-of-mind performance (Astington & Jenkins, 1999; de Villiers & Pyers, 1997). Also, exposure to discourse involving different perspectives enhances false-belief understanding (Lohmann & Tomasello, 2003).

Once the child has acquired the relevant conceptual and linguistic skills that enable explicit theory-of-mind reasoning, the child can also begin to engage in inner speech. Evidence suggests that inner speech in particular plays an important role for reflective self-consciousness. Portions of the left prefrontal lobe are associated with both inner speech and self-reflective activities, and studies using various measures of self-talk and self-reflection indicate a strong correlation between these two mental activities (Morin, 2005). According to Morin, inner speech turns the initially socially generated practice of talking and reflecting on oneself into an inner experience. As Morin points out, this idea was already expressed by Mead (1934), who argued that inner speech in early childhood serves to make young speakers aware of themselves and their individual existence through an internalization of the others' perspectives on oneself. So inner speech would reproduce social feedback and perspective taking, thereby internalizing it. Moreover, inner speech is thought to facilitate the conceptualization and labeling of self-related aspects—which can then be grouped together in a "self-file" (Perry, 2000)—thereby rendering these aspects more salient and more differentiated (Morin, 2005).

So in accordance with the conclusion of section 6.1, this section demonstrated that a number of social cognitive skills and forms of intersubjectivity are in place before the onset of linguistic abilities (indeed, these might be necessary requirements for the development of language). Nonetheless, linguistic abilities in turn seem to be necessary for the development of a full-fledged theory of mind. This might explain why chimpanzees and other great apes do not seem to be able to ascribe (false) beliefs or other mental states to others—with the probable exception of visual perceptual states—although they do seem able to engage in shared attention and selective imitation and display signs of mirror self-recognition (Call & Tomasello, 2008). I discuss the issue of self-consciousness in nonhuman animals in more detail in the next chapter.

Note that the argument is not that prelinguistic forms of social cognition, which rely on an implicit understanding of others, are being replaced or abolished by later, linguistically mediated and explicit forms of mentalizing. Rather, as explained in the previous chapter, the model I am proposing assumes that primitive forms of social cognition are retained, such that our social cognitive skills become gradually enriched and more complex as implicit information is reiteratively redescribed into more explicit formats. So we have various ways of understanding and interacting with others, some of which are based on an implicit representation of the mental states of others, and others are based on more or less explicit forms of representation. This suggests that these different forms of social cognition also require different kinds of explanation, which is an issue that I address in the next section.

## 6.4   Which Theory of Mind? Toward an Integrative Approach

The term "theory of mind" has so far been used in a general manner to refer to the ability to ascribe mental states to others and to oneself. However, there is quite a large controversy within philosophy and psychology as far as the status and nature of this theory are concerned. The most important competing theories are the theory-theory, the simulation theory, and, more recently, the interaction theory and the narrative theory (with the latter two generally combined into a hybrid model). According to theory-theorists, our mindreading abilities consist primarily in our capacity to *theorize* about the mental states of others. In contrast, according to simulation theorists, our mindreading abilities should rather be understood as the capacity to *simulate* the mental states of others. The interaction theory, in turn, posits that we possess perceptual and emotional resonance processes in early infancy, which enable us to *directly perceive* the mental states of others, and these are subsequently transformed by encounters with narratives, which accounts for the development of a more sophisticated folk psychology (Gallagher & Hutto, 2008). In this section, I discuss these different theoretical approaches in turn and argue that—insofar as there is a substantive difference between them—they should be seen as complementary rather than competing accounts.

The theory-theory assumes that mentalizing is primarily the ability to ascribe mental states to others for the purpose of predicting and explaining their behavior, and this ability relies on a folk psychological theory concerning the way the mind functions, that is, a set of rules that connects beliefs, desires, and other mental states to each other and to specific types of behavior. According to some authors, this theory is largely learned and is developed in a manner that is similar to the way scientists develop scientific theories (Gopnik & Meltzoff, 1997; Gopnik & Wellman, 1992). Other authors argue that the core of the theory is innate and hardwired, although it might require experience as a trigger (Baron-Cohen, 1995).

In contrast to the theory-theory, the simulation theory argues that it is more parsimonious to assume that, rather than having to rely on a theory of how the mind works, we simply rely on our own minds to predict and explain the behavior of others (Goldman, 1992, 2006; Gordon, 1986; Heal, 1986). According to this theory, the way we come to understand the mental states of others is by generating equivalent states in ourselves. In other words, to anticipate how someone else will behave in a certain situation, we "put ourselves in the other's shoes" and simulate how we would behave in the situation. Traditionally, both simulation theory and theory-theory agree that subjects need to be in possession of mental-state concepts to ascribe mental states to others for the purposes of predicting and explaining their behavior. However, there is also an interpretation of simulation theory according to which simulation need not imply a conceptualization of the other's mental state; we may internalize another's facial expression, for instance, without explicitly identifying the emotion it expresses. Thus what is of primary importance to a simulation theorist in this version of the theory is not that we conceptualize the mental states of others but that we embody them (Gordon, 2009; on so-called "low-level mindreading," see also Goldman, 2006).

A third, more recent alternative, interaction theory, argues that we need neither theoretical inference nor simulation to explain and predict the behavior of others, for we can directly perceive their emotions, intentions, and motives (Gallagher, 2007). And a fourth alternative is narrative theory (Hutto, 2008; Nelson, 2007), according to which the development of a theory of mind can be attributed to the narrative practices that we are engaged with. According to this theory, the child's engagement in socially guided and supported storytelling activities leads to the development of mindreading abilities. The relevant stories are said to be able to play this role because they present the ways in which the psychological attitudes of their protagonists interact with each other and with other kinds of psychological states. Being engaged with these narratives then enables children to become skillful at interpreting the behavior of others in folk psychological terms (Hutto, 2008). Because the narrative theory must assume that the child already has a basic, "practical" grasp of mental states before it can engage in storytelling activities, these last two accounts are usually defended in combination with each other:

> There is no need to appeal to standard theory-of-mind and simulative explanations of how we understand others as the basis for making sense of them folk psychologically. What begins as perceptual and emotional resonance processes in early infancy, which allow us to pick up the feelings and intentions of others from their movements, gestures, and facial expressions, feeds into the development of a more nuanced understanding of how and why people act as they do, found in our ability to frame their actions, and our own, in narrative ways. Our everyday abilities for intersubjective engagement and interaction are, in the later stages of childhood, transformed by encounters with narratives. (Gallagher & Hutto, 2008, p. 34)

Although the debate is often couched in terms that suggest that theory-theory, simulation theory, and interaction/narrative theory are competing and mutually exclusive accounts of social cognition, in the view defended here, it seems plausible that they instead describe different and complementary processes. Insofar as both theory-theory and simulation theory presuppose the possession of mental-state concepts for the ascription of beliefs and desires to explain and predict the behaviors of others, both theories seem to be incompatible with those primitive forms of social cognition that rely on an implicit, nonconceptual grasp of the other.[28] However, if simulation is instead taken to consist in a cross-modal self–other matching (whose neural basis might be found in the mirror neuron system),[29] the forms of social cognition and interaction associated with the phenomena of primary and secondary intersubjectivity described earlier do indeed seem to involve such a process of simulation and are thus compatible with this version of simulation theory. This need not necessarily be incompatible with Gallagher's interaction theory, either. The problem with that theory, as I see it, is that it says little about how we are to understand the "direct perception" of the mental states of others. Gallagher's interpretation of simulation theory seems to suggest that, similar to theory-theory, it is committed to a concept-dependent mindreading, which is why he rejects it. But, as we have just seen, not all versions of simulation theory are committed to this view. So in the broader reading of simulation theory suggested by Gordon (2009), we might interpret simulation theory as providing a (partial) account of what is going on in basic intersubjective processes, such as imitation, social referencing, and shared attention, thus explaining how we can engage in these processes based on direct perception—where the perceptual information about the other is automatically and implicitly matched with our first-person experience—without having to rely on conceptual abilities or theoretical inferences. The suggestion is that we are in a position to "pick up the feelings and intentions of others from their movements, gestures, and facial expressions" (Gallagher & Hutto, 2008, p. 34) precisely because the perception of these movements, gestures, and facial expressions elicits in ourselves feelings and intentions that are similar to the ones expressed by the other. Importantly, as I argued earlier, at this level we neither explicitly distinguish between self and other nor engage in the explicit attribution of mental states to self or other.

The ability to differentially ascribe mental states to self and other requires redescribing the information implicit in these forms of basic social interaction into a more explicit format. Thus the kind of theorizing presumed by a theory-theorist, which would begin as the self–other matching that is implicit in primary and secondary intersubjectivity, becomes redescribed into a more explicit format, which then allows for a distinction between self and other and, ultimately, for the explicit ascription of mental states to self and other. According to the theory of representational redescription, we would first observe something like a "theory-in-action," where the behavior of the child indicates that it uses an explicit (albeit not yet conceptual or linguistic) representation of the mental states of others to explain and predict their behavior. Indeed, something like this seems to be happening in cognitive empathy, where children learn to distinguish their own emotional experience from that of another so as to console the other person, or when they begin to understand that others have a different perspective from themselves. The precursors of this might also be what underwrites the early mindreading abilities found in the studies by Onishi and Baillargeon (2005) and Buttelmann et al. (2009).

Finally, as we reach the level of explicit mentalizing involved in the mastery of explicit false-belief tasks, we can speak of a true theory of mind in the theory-theory sense, which is used to provide rational justifications for the behavior of others and of oneself in terms of the belief-desire structure that is familiar in folk psychological explanations. At this stage, the distinction between self and other, as well as between propositional attitude and propositional content, is explicitly represented (in what Karmiloff-Smith calls E2/3 format). So it is at this stage that children begin to appreciate the representational nature of mental states and to use this understanding to account for the behavior of others, thus warranting the talk of a theory of mind in the theory-theoretical sense.

Moreover, it seems plausible that the ability to theorize about mental states is directly influenced, and perhaps even enhanced, by the narrative practices that the child is embedded in. That is to say, once the child begins to speak and to access representations of the intentional relations of self and other in an explicit, conceptual, and linguistic format, these representations will be shaped by the social narratives that the child is engaged with. Presumably, these socially generated narratives can also be turned into inner speech and thus internalized, thereby influencing processes of self-reflection and shaping the developing self-concept. Thus, in my view, we should see narrative theory not as an alternative to theory-theory but rather as describing a complementary process that modifies and shapes how we theorize about ourselves and others.[30]

In short, I am suggesting that simulation theory, theory-theory, interaction theory, and narrative theory all have the potential to contribute to a better understanding of social cognition, and they are therefore best seen as complementary rather than competing theories.[31] That is to say, we should aim for a pluralist and integrative approach to social cognition. As pointed out earlier, intersubjectivity is a multifaceted phenomenon that requires multifaceted explanatory strategies. Crucially, this is true not just for a developmental account of intersubjectivity but also for an account of social cognition in mature adults. As explained earlier, it is not the case that the kind of low-level basic processes involved in early stages of social cognition become replaced or abolished by later, more explicit levels of mentalizing. Rather, even as mature adults, we have various ways of understanding and interacting with others that exist alongside each other and require various kinds of explanation.

Notice that none of the foregoing discussion excludes the possibility that, in fact, we do not need to appeal to the mental states of others for many of our day-to-day social interactions. I have focused on mindreading abilities—and have proposed a pluralist and integrative approach to account for them—because my thesis was that it is only by contrasting the mental states of others with one's own that one becomes aware of one's own mental states as such. But this is compatible with the claim that in many situations of social interaction, we appeal not to others' beliefs and desires but to their environment, personality, social role, cultural norms, or past behavior (see Andrews, 2012). Indeed, I agree with Andrews that a satisfying theory of social cognition should incorporate the broad range of factors that play a role in social interaction, rather than focusing solely on the attribution of propositional attitudes.[32] Nor do I take the claims I have made here to be incompatible with the thesis that our distinctively human capacities for mindreading might not have evolved without our capacities for "mindshaping," such as those involved in conformism, or norm institution and enforcement (Zawidzki, 2013).

### 6.5 The Parallelism between Understanding Self and Other in Normal Development and Autism

According to my proposal, explicit self-representation develops in parallel with the representation of others as subjects with goal-directed, epistemic, and emotional intentional relations. The recognition that others are intentional agents like oneself leads to a contrastive differentiation between the intentional states of others and one's own intentional states, and thus to self-consciousness in the sense of having the ability to think "I"-thoughts.

In addition to the empirical evidence discussed so far, which provided evidence for the claim that the ability to ascribe intentional relations to others and to oneself develops via a reiterative process of representational redescription, further empirical studies directly address the correlative development of self- and other-representations and speak in favor of this thesis. We have seen already that at the first level of explicit representation, studies have shown correlations between mirror self-recognition and empathy, thus providing evidence for the thesis that self-consciousness develops in parallel with the understanding of others (e.g., Bischof-Köhler, 1988). Likewise, recent empirical studies suggest a parallelism between self- and other-representation at the level of the possession of an explicit theory of mind. The relevant empirical evidence must be regarded as preliminary, as the majority of research on theory of mind has focused on the problem of "reading other minds" rather than the self-attribution of mental states; however, several studies do seem to confirm such a parallel development (for a review, see Happé, 2003). Most impressively, a meta-analysis of theory-of-mind studies (involving 178 separate studies) concludes that children pass self-belief tasks at the same time as they pass other-belief tasks:

> The essential age trajectory for tasks requiring judgments of someone else's false belief is paralleled by an identical age trajectory for children's judgments of their own false beliefs. Young children, for example, are just as incorrect at attributing a false belief to themselves as they are at attributing it to others. (Wellman, Cross, & Watson, 2001, p. 665)

This confirms that there is a parallel development for the explicit representation of one's own mental states and those of others.

Research on autism seems to further support this parallelism, although I do not want to emphasize this kind of research too much here. For one thing, we should generally be careful in extending findings obtained through the study of disorders and abnormal developments to general principles. This is especially so when the disorder, as is the case for autism, is a spectrum disorder, so that individual results and studies are difficult to compare. Moreover, although much progress has been made in understanding this disorder in recent years, the fact remains that the basic mechanisms underlying it are still for the most part ill understood. Nonetheless I would like to present some relevant findings from the literature on autism here, bearing in mind that they should be treated with caution.

Generally, autism is thought to be a disorder that affects mentalizing abilities (Baron-Cohen, Leslie, & Frith, 1985; Frith, 2003). Whether this is due to a deficit in abilities that are specific to mindreading, or whether its basis lies in the disruption of more general processes, is controversial. For instance, in contrast to normally developing children, children with autism do not show a preferential attention bias for social stimuli. They show a general tendency to avoid eye contact and do not display shared attention or proto-declarative communicative behaviors. It is possible that the later difficulties in mindreading tasks displayed by children with autism have their origin in their lack of exposure to relevant social stimuli due to their earlier avoidance behavior. Other hypotheses regarding the causes of impairments in mentalizing abilities include the executive function hypothesis, which posits an impairment of higher-level attentional and control mechanisms in people with autism, and the central coherence hypothesis, which suggests that people with autism process information in a way that tends to focus on detail and disregard context (Frith, 2003).

Nevertheless, in congruence with the evidence concerning normally developing children cited earlier, a number of studies involving children with autism have shown that their ability to attribute mental states to others is correlated with their ability to ascribe mental states to themselves. For instance, similar to normally developing children under the age of four, children with autism show no advantage for reporting on their own knowledge versus the knowledge of others (Kazak, Collis, & Lewis, 1997; Perner et al., 1989) and are unable to report on their initial intention when the outcome of an action conflicts with the original intention (Phillips, Baron-Cohen, & Rutter, 1998). Moreover, a study by Frith and Happé (1999) testing three adult subjects with Asperger's syndrome using the "descriptive experience sampling method" developed by Russell Hurlburt has shown that the level of performance in theory-of-mind tasks appears to be positively correlated with the ability to engage in introspection.[33] That is to say, to the extent that people with autism have difficulties "reading" the mental states of others, they also seem to have difficulties in metarepresenting their own mental states as mental states. Finally, a recent study by Williams and Happé (2009) using an unexpected-content task suggests that children with autism might even have greater difficulty representing their own beliefs than the beliefs of others. They found that participants with autism spectrum disorder, in contrast to typically developing children, found it significantly more difficult to report their own prior false belief than to predict the false belief of another person on the "plasters task."[34]

However, it is not the case that children with autism—at least those who are on the milder end of the spectrum—cannot develop a theory of mind at all. Yet to the extent that they do acquire a theory of mind, they seem to need to rely on compensatory learning mechanisms for the attribution of mental states. So their mentalizing abilities lack the automaticity required for normal social interactions, and the way they acquire these abilities differs from that seen in normally developing children (Frith, 2003). This interpretation is confirmed by a recent neuroimaging study by Dapretto et al. (2005), which found that although children with autism were as able to imitate emotional expressions of others as normally developing children, the means they used to master this task appear to be quite different. In contrast to normally developing children, their mirror neuron premotor and insula areas were not as active, and the degree of involvement of these areas during imitation was inversely related to the severity of their diagnosis of autism. Instead, other brain areas appeared to be activated, suggesting an alternative route to imitation (for a more detailed discussion, see Barresi & Moore, 2008). So this study seems to suggest that, in contrast to normally developing children, children with autism cannot rely on mirroring (i.e., embodied simulation) in the early stages of theory-of-mind development. Rather, they seem to have to go exclusively via the theory-theory route (perhaps complemented with narrative practices), relying on observation complemented with explicit "rules" and inferences for the attribution of mental states.

A possible implication of this finding could be that people with autism have difficulty matching and integrating their first-person experiences of their own mental states with the third-person information they receive about the behaviors of others (Barresi & Moore, 2008). This would make it difficult for them to understand mental phenomena, because they are unable to form "a direct connection between the two necessary, inseparably tied aspects of all mental phenomena, an externally available bodily expressive component, and an internally available feeling component" (Barresi & Moore, 2008, p. 61). As a consequence, insofar as autistic subjects do reflect on the mental lives of themselves and others, they might develop rather different accounts: on the one hand, they might develop complex theoretical accounts from a third-person view; on the other hand, they might tend to overgeneralize from their own first-person perspective. Thus they could end up deploying independent first-person and third-person theories of mind, resulting in a "naive egocentrism" on the one hand, and a highly abstract social understanding on the other hand, combined with an inability to easily switch between the two (Frith & de Vignemont, 2005).

## 6.6 Summary and Conclusion

This chapter has shown how explicit self-representation requires an awareness of other (embodied) subjects and of their similarity to oneself, such that one can contrast one's own bodily and mental states with those of others and come to think of oneself as a subject among others. This awareness develops over the course of an increasingly complex process of perspectival differentiation, during which information about self and others that is implicit in early forms of social interaction becomes redescribed into an explicit format.

This level-based account of self-consciousness and intersubjectivity can be roughly summarized in the following schema (table 6.1). Notice that while based on the distinctions introduced by Karmiloff-Smith (see previous chapter), the schema presented here also differs from her framework

**Table 6.1**

Levels of self- and other-representation

| Level | Age | Social cognitive abilities | Representational format | Cognitive mechanism (?) |
|-------|-----|---------------------------|------------------------|------------------------|
| 0 | Birth onward | Neonatal imitation, detection of multimodal contingencies | Automatic cross-modal matching, no self–other differentiation | Mirroring |
| 1 | 9–12 months onward | Shared attention, social referencing, proto-imperatives and proto-declaratives | Implicit representation of self–other and of intentional relations, domain specific | "Direct perception" (i.e., embodied simulation) |
| 2 | 14–18 months onward | Selective imitation, mirror self-recognition, self-conscious emotions, empathy, perspective taking (level 1), implicit false-belief tasks | Explicit self–other differentiation and representation of intentional relations, but no conceptual representation of mental states as such | Simulation + "theory-of-mind-in-action" + narrative practice |
| 3 | 4 years onward | Mastery of explicit false-belief and unexpected-content tasks, understanding of perspectives (level 2), appearance–reality distinction, intentional deception | Explicit (conceptual) representation of propositional attitudes | Explicit mentalizing (theory-theory) + narrative practice |

in that it posits an additional, even more basic level (which makes no representational distinction, whether implicit or explicit, of self and other). Notice also that the schema is only a rough sketch of the complex processes that were discussed in this chapter; the different levels should not be seen as being strictly separated from one another but rather as gradual transitions. There are no strict boundaries between the levels, and the representational redescription of implicit information into more explicit formats may occur at different speeds and times throughout different domains. At any given age, a child may already possess higher-level cognitive abilities in one domain while still being at a lower level in another domain. Finally, again, the claim is not that lower-level representations are replaced or abolished once a higher level of representation is reached. Rather, the schema is meant to illustrate a gradual enrichment of various ways of representing self and other, which persist into adulthood.

So the account presented here suggests a gradual transition from implicit to explicit forms of self- and other-representation that leads to an increasingly complex array of social cognitive abilities, which in turn lead to the development of a self-concept. We can now see how we get from the self-related information that is implicit in perception and bodily forms of self-awareness to an explicit representation of oneself. The crucial element is intersubjectivity, which requires a mechanism that enables the matching of first- and third-person information in concert with a process of representational redescription.

We have also seen that we have multiple ways of understanding and engaging with others, which require multiple explanatory strategies. Some of these are likely to involve simulation processes and rely on bodily and implicit self–other matching; others require explicit propositional attitude ascriptions and linguistic abilities. Once the level of conceptual and linguistic self–other representation is reached, communicative actions, as well as personal and cultural narratives, can begin to shape an individual's self-notion and influence her self-reflection and subsequently self-conceptualization as belonging to particular groups or cultures.

In the next chapter, I examine the implications of these results for the question of whether—and at what level—we can attribute self-consciousness to nonhuman animals by applying the schema outlined here to several cognitive abilities associated with self-consciousness that are found in nonhuman animals.

# 7 Self-Consciousness in Nonhuman Animals

What are the implications of the account I have presented for the question of whether nonhuman animals can be self-conscious? As we will see, the answer to this question is complex and at present must remain to some extent undecided. Part of the problem is that, to study self-consciousness in nonhuman animals, researchers have to develop paradigms that do not rely on linguistic abilities yet probe cognitive abilities that are clearly associated with self-consciousness and produce unambiguous results. This can be extremely difficult, if not impossible, to achieve, as we can see when trying to interpret studies of both prelinguistic humans and nonlinguistic nonhuman animals.

So far, the study of self-consciousness in nonhuman animals has proceeded mainly by focusing on three different abilities: the ability for mirror self-recognition; the ability for mindreading; and the ability for metacognition. I will discuss these in turn. As we will see, it is not always obvious to what extent these abilities indicate self-consciousness. Moreover, the experiments in question often remain inconclusive because the paradigms employed are unable—either in principle or due to current limitations—to produce unambiguous evidence for the abilities under investigation. Nonetheless I believe we can gain some important insights from these experiments, and the multilevel framework developed in the previous chapter can help in getting a more specific answer to the question of what kind(s) of self-awareness we can ascribe to nonhuman animals. Future research will no doubt add to the picture that emerges when we combine the insights gained in the previous chapters with the empirical evidence available to date.

## 7.1   Mirror Self-Recognition

As we saw in the previous chapter, the ability to recognize oneself in the mirror is often taken to be an indicator of self-awareness (Gallup, 1979; Anderson & Gallup, 1999). It has long been argued that chimpanzees and orangutans possess the ability for mirror self-recognition (e.g., Suarez & Gallup, 1981), and this ability has also been attributed to dolphins (Marten & Psarakos, 1994) and, more recently, elephants (Plotnik et al., 2006) and magpies (Prior, Schwarz, & Güntürkün, 2008).[1] However, as discussed previously, the ability to use a mirror to perform self-directed actions, such as the removal of a mark from parts of the body that are otherwise visually inaccessible, does not necessarily imply the ability to *recognize* oneself in the mirror. Rather, the ability to perform such self-directed actions can be explained in terms of an animal's ability to use "novel, displaced visual feedback on the animal's physical state and behavior" (Heyes, 1994, p. 916). The visual information that is received from the mirror is contingent on the physical state and movements of the animal; hence it is a form of feedback. While it is a novel way of gaining such feedback in that it is "displaced" (i.e., not continuous with the animal's body), the feedback as such is not different in kind from the feedback an animal would receive by observing parts of its body without a mirror. Thus, according to Heyes, while the ability to pass the mirror test requires the ability to map such visually received information to simultaneously received information through proprioception, and to use such a mapping to guide one's actions (e.g., to remove a mark placed on an otherwise inaccessible part of the body), it does not in and of itself imply the ability to recognize oneself or to explicitly distinguish oneself from others. Put differently, to pass the mirror test, an animal must arguably be capable of "kinesthetic-visual matching" (Mitchell, 1993, 2002), or of collating different representations of the same object (Suddendorf & Butler, 2013), but this does not imply an awareness of oneself as distinct from others or as capable of being perceived by others. So the ability to pass the mirror test in and of itself does not suffice to demonstrate self-awareness (and if it were, it would at most demonstrate awareness of one's physical appearance); rather, to indicate self-awareness, it would need to be accompanied by other social cognitive abilities that demonstrate an ability to explicitly distinguish between self and other (and an understanding that one can be perceived by another).

Earlier I argued that insofar as the ability to pass the mirror test is found to be accompanied by other phenomena, such as a display of coyness or embarrassment, as well as social cognitive abilities, such as level 1 perspective taking or empathy, inference to the best explanation might well lead us to ascribe a basic form of self-awareness (i.e., self-awareness at what I have termed level 2 in the schema presented in the previous chapter) to children who pass the mirror test. Naturally the same principle applies to nonhuman animals. To my knowledge, we have no evidence to date that mirror self-recognition in animals is accompanied by displays of embarrassment or similar phenomena. That said, there does seem to be some evidence for empathy and other social cognitive abilities in animals that pass the mirror test, including basic forms of mindreading, though this remains somewhat controversial (see sec. 7.2 for further discussion). Hence the ability to pass the mirror test does not present unambiguous evidence to suggest the presence of basic forms of self-awareness in nonhuman animals.

Assuming that nonhuman animals are able to recognize themselves in the mirror, as well as display other social cognitive abilities that indicate the ability to differentiate between self and other, would imply self-awareness at what I have labeled level 2 in my framework. However, it would not yet imply the presence of a self-concept in the sense of an explicit conceptual distinction between self and other, as it would not imply the ability to explicitly attribute mental states as such to oneself and to others. Thus at best the ability for mirror self-recognition—when taken together with other social cognitive abilities—suggests a sense of self at what I have termed level 2 (or what would be a level E1 representation in Karmiloff-Smith's terminology).

## 7.2 Mindreading

So what do we know about mindreading abilities in nonhuman animals? In the previous chapter, I argued that self-consciousness requires the ability to distinguish one's own mental and bodily states from those of others. Accordingly, the ability for mindreading is a central aspect of self-consciousness. Indeed, in the view defended here, self-consciousness and intersubjectivity are two sides of the same coin. To what extent do nonhuman animals possess mindreading abilities, and thus self-consciousness? Some evidence indicates that chimpanzees are capable of understanding others' goals (reviewed in Call & Tomasello, 2008). For example, chimpanzees show different reactions to a human who is unable to give food compared to a human who is unwilling to, and to accidental versus purposeful behaviors. They also seem able to discern another's goal being out of reach and either help them obtain it or grab the food a competitor is trying to reach. Moreover, they have been shown to engage in gaze following and instrumental gestural communication in a manner similar to that of human infants, and they seem able to engage in selective imitation (see previous chapter for a discussion of selective imitation). In addition, in competitive situations, chimpanzees seem able to take account of what another chimpanzee can or cannot see or hear—that is, they can engage in level 1 perspective taking— and they can keep track of what another has just seen a moment earlier. Similar results have been obtained in rhesus monkeys (Flombaum & Santos, 2005; Santos et al., 2006) and scrub jays (Dally et al. 2006; Emery & Clayton, 2001; Clayton et al., 2007). While most of the experimental paradigms that were used in these different studies cannot rule out a behavioral-rule interpretation, as Call and Tomasello (2008) point out, the behavioral rule in question would have to be different in each case, so that it would seem to be more parsimonious overall to assume that chimpanzees (and possibly monkeys, as well as some birds) are able to attribute goals and perceptual states to others to account for the relatively large number of different experimental results to this effect.[2]

However, we have no conclusive evidence that chimpanzees are able to attribute mental states beyond intentions and perceptual states (Call & Tomasello, 2008). Indeed, experiments that were intended to test for the ability of false-belief attribution (as compared to the attribution of perceptual mental states) produced negative results (e.g., Call & Tomasello, 1999; Kaminski, Call, & Tomasello, 2008). That said, it is a well-known insight that the absence of evidence is not to be confused with evidence of absence. Therefore it cannot be ruled out that future studies will produce evidence in favor of the ability for belief attribution in chimpanzees and other animals. Already some evidence suggests that dolphins are able to attribute (false) beliefs to others (Tschudin, 2006), though further studies are required to replicate these results.

Thus, given the evidence to date, it seems that chimpanzees, as well as some other animals, almost certainly possess the ability to implicitly represent the intentional relations of others (level 1) and possibly the ability for an explicit self–other differentiation at level 2. This situates their social cognitive skills somewhere in between the abilities of nine- to twenty-four-month-old infants, consistent with the fact that the ability to pass the mirror test is also shared by chimpanzees (and possibly some other animals) and infants at about eighteen to twenty-four months. Accordingly, chimpanzees, and possibly some other animals, seem to possess self-awareness, as well as awareness of other subjects, at least in a basic sense. However, based on present evidence, they cannot be attributed with a self-concept in the sense of an explicit representation of the self and a grasp of mental states as such, that is with level 3 self-awareness (with the possible exception of dolphins).

## 7.3   Metacognition

Recently the study of self-awareness in nonhuman animals has been expanded to include the study of so-called metacognitive abilities. Metacognition is often defined as "thinking about thinking," or as the ability to represent and self-attribute one's own mental states, as in the explicit thought "I know that x" or "I feel uncertain about x." It is thus regarded as a form both of metarepresentation and of self-consciousness. Recent experiments have tested metacognitive abilities with the help of nonverbal tasks, which makes it possible to study metacognition in nonhuman animals. These experiments generally involve stimulus-discrimination tasks with varying degrees of difficulty. They proceed by presenting the subject with a set of stimuli, for example, high-pitch versus low-pitch sounds, or densely versus sparsely pixelated visual displays, that have to be distinguished from each other. Some discriminations are relatively easy, whereas others are difficult. In any particular trial, subjects have the option either to respond or to opt out from responding on the current trial. They will usually receive an attractive reward for giving the correct response and a somewhat less attractive reward (or a "time-out") for choosing to opt out. Studies find that monkeys, as well as dolphins, show a pattern of response that is very similar to that of humans, which is to say that they are able to effectively use the "opt-out" button in trials that are perceived to be difficult (Smith et al., 1995; Shields et al., 1997). This pattern is often interpreted as showing that humans as well as nonhuman animals are aware of how much they know (i.e., how confident or certain they feel about being able to give a correct response), and hence they possess self-consciousness with respect to their state of knowledge.

However, the experiments in question do not have to be interpreted in this fashion. An alternative interpretation would be that rather than demonstrating the ability to represent one's own mental state, the experiments test the ability to represent a third class of stimulus categorization. So, for example, if the task consists in judging whether auditory stimuli are high (which should be followed by pressing the "high" button) or low (which should be followed by pressing the "low" button), the opt-out button could represent a third category, such as "middle." In other words, the response behaviors in tasks of the sort just described could simply be based on "object-level processes," rather than being based on a feeling of uncertainty (Son & Kornell, 2005).

An objection to this explanation is that experiments with human participants have shown that subjects often explain their pattern of response by referring to their own mental states and report that they chose the opt-out choice for trials in which they felt uncertain. Although animals cannot give verbal explanations of this kind, the similarity in response pattern could suggest that a similar mechanism is at work (Smith, 2005). Moreover, in response to the concern that the opt-out button represents a third category, rather than indicating metacognitive awareness of one's state of uncertainty, alternative tasks have been developed that are based on memory rather than on the concurrent presentation of the stimulus. For example, monkeys were shown to be able to master a task in which they were presented with an icon that they would later, after a delay, have to recognize among an array of other items (Hampton, 2001). Crucially, the monkeys had to decide whether to take the test before being presented with the test stimuli. Alternatively, they could opt for an easier task. It is important to note that in this case the monkeys cannot rely on "objective uncertainty"; they must rely on a prospective judgment of their memory, that is, on a "feeling of knowing" (or feeling of uncertainty), to solve the task, for they have not yet seen the test stimulus (Hampton, 2005). In addition, Son and Kornell (2005) developed a "betting" task in which monkeys first had to perform a discriminative task and then, once the stimulus disappeared, make a high- or low-confidence judgment. Again, in these tasks, the objective features of the stimulus are separated from the subjective "feeling of knowing/uncertainty."

Still, Carruthers (2008) has argued that even in these tasks, the mental state that is driving the animals' behavior (e.g., that is making them either opt out or take the test) does not have to be interpreted as a metarepresentational state, that is, a mental state that is directed at another mental state. He suggests that the behavior in question can be explained as a "gating mechanism" that is based on a combination of various mental states entertained by the animal—in particular a combination of one weak belief (e.g., the belief that the pattern is dense) with another equally weak belief (e.g., the belief that the pattern is sparse) and with the desire to obtain food or to avoid a time-out— without any of these states having to be higher-order states. Moreover, Kornell (2014) has recently pointed out that humans often seem to rely on internal cues such as ease of processing or familiarity, rather than on a direct access of their memory, when making metacognitive judgments.[3] For example, numerous studies have shown that presenting words in a larger font or priming certain stimuli affects judgments of certainty, even when there is no correlation between these manipulations and the actual strength of memory (for a review of relevant studies, see Kornell, 2014). Given that the paradigms used in the study of animal metacognition are comparable to the paradigms used in these studies, it seems plausible that animals, like humans, rely on cues such as ease of processing or familiarity when making metacognitive judgments, rather than directly accessing their memory states (Kornell, 2014). This suggests that "feelings of knowing" (or "feelings of uncertainty"), do not imply direct access to memory.[4] In addition, these feelings can often remain unconscious.

Finally, even if the behavior shown in studies of metacognition were to express a conscious "feeling of knowing" (or "feeling of uncertainty"), and even if such a feeling did suggest access to a lower-level mental state, this would imply neither that the feeling is presented *as being directed at another mental state* nor that it is presented as representing *one's own* state of memory or knowledge. Although human participants, who possess the self-concept, will upon explicit reflection verbalize and report this feeling by means of an explicit self-ascription (e.g., "I feel certain that I remember this"), this is not to say that the experience of the feeling itself, and the ability to exploit this feeling for behavioral control, already imply an explicit awareness of having access to a lower-level mental state, or of oneself as the subject of the first-order mental state. Put differently, perhaps it is reasonable to assume that the behavior of nonhuman animals in experiments that study metacognition is indicative of them experiencing a certain mental state, which would be implicitly directed at another mental state, in the sense of *de facto* containing information about the lower-order state. For instance, the animals might be experiencing a state of surprise, indicating a conflict between different representational states (Proust, 2006). Even so, this does not imply that the animals in question are aware of the feeling *as* giving them information about another mental state of theirs. So metacognition in this sense is not a form of explicit metarepresentation (cf. Proust, 2006). Nor do we have reason to think that the animals in question can explicitly distinguish between themselves and their own mental states and those of others. Accordingly, we have no reason to think that they would be able to explicitly self-ascribe a feeling of surprise or uncertainty.[5]

Similar to the mirror test case, we would only have some (albeit indirect and inconclusive) reason to make such an assumption if metacognitive abilities in nonhuman animals were reliably correlated with social cognitive abilities that suggest the ability to explicitly represent one's own mental states and the mental states of others. As we saw in the previous section, there is some, albeit controversial, evidence that dolphins—who routinely pass metacognitive tests and have also been attributed with the ability to pass the mirror test—also pass false-belief tasks and hence might possess a theory of mind (at what would be level 2 or even level 3 in my framework).[6] On the other hand, monkeys, who likewise routinely pass tests for metacognition, at the same time fail to display signs either of mirror self-recognition or of the ability to pass false-belief tasks. That said, as we saw in the previous section, rhesus monkeys have been shown to be able to take into account what another can or cannot see, and so display signs that might suggest a partial grasp of the mental states of others (at level 1 or 2).

Thus studies of metacognitive abilities, while certainly fascinating in their own right, do not in and of themselves demonstrate self-consciousness in the sense of the ability to think "I"-thoughts. Their results can be accounted for without assuming that the animals in question possess the ability to self-ascribe mental states, and thus without assuming that animals who show metacognitive abilities are self-aware. That said, as was discussed in chapter 5, metacognition might be seen as a necessary prerequisite for the representational redescription of implicit, procedural information into more explicit forms of representation, and, as such, a necessary precursor to the ability to explicitly represent and self-ascribe mental states. In fact, the feeling of uncertainty or surprise might be just the kind of emotion required to engage in processes that eventually lead to representational redescription (cf. Proust, 2006).

## 7.4 Conclusion

In conclusion, the evidence to date suggests at best that nonhuman animals possess some basic forms of self-awareness, similar to those found in human infants. Based on findings from experiments testing the ability to pass the mirror test in conjunction with those testing the ability for mindreading and metacognition, it seems reasonable to conclude that chimpanzees possess a sense of self at level 2, which is to say that they can differentiate between themselves and others and ascribe basic intentional relations, such as perceptual states, but without an explicit understanding of the nature of mental states as such (i.e., as states that can misrepresent). The experimental findings are less clear regarding other species. For example, rhesus monkeys and scrub jays seem to be able to attribute perceptual states to others, and monkeys also show metacognitive abilities but they do not show evidence for mirror self-recognition. Other animals, such as magpies, elephants, and dolphins, on the other hand, show signs of mirror self-recognition and metacognition, but we don't know whether they also possess the ability to attribute mental states to others and to themselves, though there is some positive evidence in the case of dolphins. Thus we need more research to provide a more comprehensive picture of the self-representational abilities in different nonhuman animals. In particular, species that seem to perform well in certain tasks indicative of self-representation should be tested on other tasks to corroborate these findings, for, as we have seen, the ability to pass the mirror test or the display of metacognitive abilities on their own are of limited value with respect to the question of self-consciousness. Ideally, the tasks in question should test the ability to ascribe mental states both to others and to oneself.

Moreover, as Lurz (2011) argues, many of the experimental paradigms in use today suffer from methodological limitations because they cannot rule out behavioral-rule interpretations.[7] While I agree with Call and Tomasello (2008) that providing such an interpretation in each case seems ad hoc, ideally, researchers should develop experimental paradigms that could rule out that the animals in question rely on behavior reading, rather than on mental-state attributions. Lurz (2011) outlines possible paradigms that fulfill these criteria. Further, if Andrews (2012) is right, to find out whether animals are indeed able to attribute mental states to others (and to themselves), it might be necessary to shift the focus of experimental paradigms away from situations in which the behavior of another needs to be predicted and toward situations that call for the *explanation* of behavior instead.

For now, we can conclude that some nonhuman animals—in particular chimpanzees—do seem to possess at least basic forms of self-awareness (i.e., at level 2), and it seems likely that this may be true for other animals as well, though further research is needed to provide conclusive evidence and to show what forms of self-awareness exactly can be attributed to other nonhuman animals. Notice also that we cannot rule out that nonhuman animals—in particular dolphins—also possess higher forms of self-awareness (i.e., at level 3). At the same time, there is little reason to think that we should find the same representational abilities in all species. After all, it seems plausible that representational abilities, including the ability to represent oneself, occur in different degrees not just with respect to ontogeny but also with respect to phylogeny. In other words, it is reasonable to assume that different organisms are likely to vary with respect to the sophistication of their representational abilities, including the degree to which these representations can be said to express self-awareness (Van Gulick, 2006). Moreover, as I argued in the previous chapter, while a number of social cognitive skills and forms of intersubjectivity seem to exist before the onset and independent of linguistic abilities, there are reasons to think that language is necessary to develop a full-fledged theory of mind. Thus I suspect that further research is likely to produce evidence of just such varying degrees of self-awareness in different species.

# 8 Conclusion and Questions for Future Research

## 8.1 Putting the Pieces Together

Let us take stock. We started this investigation in chapter 1 with a definition of self-consciousness in terms of the ability to think "I"-thoughts, and with an analysis of traditional models of self-consciousness. These are based on the idea that self-consciousness is the result of a process of self-reflection or self-perception in the course of which a subject takes herself as an intentional object and thus becomes aware of herself. I argued that these models fail because they either have to presuppose what they attempt to explain (i.e., self-consciousness) or lead into an infinite regress. For self-reflection to bring about self-awareness, the subject needs to be in possession of self-identifying criteria that can help her to identify herself (or a given mental or bodily state of hers). In other words, the subject already needs to possess self-knowledge, namely, knowledge of the relevant self-identifying criteria. Now, either the possession of these criteria is simply presupposed, in which case the model is circular, or it relies on yet another process of self-reflection, which leads into a regress. Thus self-consciousness cannot be based on self-reflection or self-perception. As a result, some authors have argued that self-consciousness must ultimately rely on some kind of prereflective self-awareness, or sense of mineness. However, this raises the question of how this sense of mineness is to be analyzed. Alternatively, other authors have suggested that a solution to the problem of self-consciousness lies in turning to an analysis of the linguistic expression of canonical forms of self-consciousness, that is to say, by turning to an analysis of the linguistics of the first-person pronoun. This analysis brings to the forefront important characteristics of self-conscious thoughts, such as (1) their nonaccidental (i.e., nonidentificational) self-reference, (2) their immediate implication for action, and (3) their immunity to error through misidentification. Nonetheless the linguistic turn poses problems of its own, for it leaves open how the ability to refer to oneself by means of the first-person pronoun is to be explained. That is to say, it leaves unexplained what the underlying representational structure of "I"-thoughts is, such that they allow for nonidentificational self-reference, and how the ability to refer to oneself in thought can develop.

The next step in the investigation was to see whether theories of nonconceptual content can help to solve the problem of self-consciousness. It has been suggested that some forms of self-consciousness (i.e., some forms of "I"-thoughts) are independent of conceptual and linguistic abilities and can provide the basis for more complex forms of self-representation. To see whether this claim can withstand scrutiny, we first turned to an analysis of nonconceptual content in general (chap. 2). It turned out that there are good arguments in favor of the notion of nonconceptual content; that is, there are forms of representation that are best described as being nonconceptual. I also argued that the structure of nonconceptual representations is such that they are noncompositional (or structure-implicit) and do not allow for context-independent general thought, and their content is determined (at least in part) by the organism's abilities to interact with the environment. In other words, nonconceptual content is best thought of as a form of "knowing-how" rather than as a form of "knowing-that."

In chapters 3 and 4, this analysis was put to work to investigate theories of so-called nonconceptual self-consciousness. One can distinguish between self-representationalist theories of nonconceptual self-consciousness and non-self-representationalist (or "no-self") theories. According to self-representationalist theories, there are nonconceptual forms of representation, such as ecological perception and bodily experience, that (1) are forms of *self*-representation (i.e., their content contains an explicit subject component); (2) have immediate implications for action; and (3) are immune to error through misidentification. Hence, in this view, these nonconceptual forms of representation are thought to fulfill all three criteria for self-conscious thought and should therefore be regarded as forms of self-consciousness, even though they do not require the ability to apply a self-concept or to possess mastery of the first-person pronoun. If so, these nonconceptual forms of self-consciousness can provide the foundation for other, more complex and conceptual forms of self-representation.

However, although it is true that perception and bodily experience contain implicitly self-related information and have immediate implications for action, they do not explicitly represent the self and hence do not fall under the category of thought to which the immunity principle can apply. This is because there is a difference between implicitly self-related (i.e., agent-relative) information and explicit self-representation. The former is sufficient for action guidance, but the latter is required for self-reference. Moreover, because states with nonconceptual content do not contain a self-referring component, they cannot be said to be immune to error through misidentification. The immunity principle requires the possibility of misrepresenting the property that is self-ascribed while correctly representing the subject. In other words, a self-conscious thought that is immune to error through misidentification can "what"-misrepresent, but it cannot "who"-misrepresent. However, if nonconceptual content does not represent the subject, then the question of whether it can "what"-misrepresent while correctly representing the "who" cannot even be sensibly applied. Immunity is a property of judgments, not of the perceptual representations that might ground these judgments. Alternatively, if the self-representationalist was right that the subject is represented in experience, it could not be immune to error through misidentification because nonconceptual content is noncompositional. In this case, a misrepresentation of the property would entail a misrepresentation of the subject as well.

Thus self-representationalist theories of nonconceptual self-consciousness are misguided, because they fail to distinguish between implicitly self-related information and explicit self-representation. What we need, therefore, is a non-self-representationalist theory (chap. 4). According to such a theory, it is precisely because the self is not explicitly represented in perception and bodily experience that these states can provide the basis for explicit "I"-thoughts or *de se* thoughts that are immune to error through misidentification. This is because—in cases where the "I" is used "as subject"—the "I"-thought only makes explicit what was already implicit in the mode of presentation. I also argued that non-self-representationalist theories can provide us with the kind of analysis of prereflective self-consciousness that was found missing—and that some authors have claimed to be unobtainable.

Nonetheless, to solve the problem of self-consciousness, we still need an account that can make intelligible the transition from implicitly self-related information to explicit self-representation. In other words, we need an account of how the nonconceptual content of experience becomes conceptualized, thus allowing for the formation of explicit "I"-thoughts. To provide such an account, chapter 5 spelled out the general distinction between implicit and explicit representation in more detail and presented a model of the transformation from one to the other via a process of representational redescription. It turned out that the distinction between implicit and explicit representation does not map easily onto the distinction between nonconceptual and conceptual content, as it transcends the dichotomy that is implied by the latter. This shows that it is necessary to distinguish between different level of explicitness, and the transition from implicit to explicit representation occurs via a reiterative process of representational redescription that leads from context-dependent, special-purpose procedural representations via "theories-in-action" to conceptual, general, and systematically decomposable and recombinable representations. I also argued that mental representations consist of different components, each of which can be either implicitly or explicitly represented. For example, it is possible to have an explicit representation of a propositional content, while the propositional attitude or the holder of the attitude is only implicitly represented.

Chapter 6 applied the results of this general analysis to the problem of self-consciousness. I argued that self-consciousness develops through a process of an increasingly complex differentiation between self and other, which ultimately leads to an explicit theory of mind that can be applied to others and to oneself. It is only when one realizes that certain representational states can be had by others—that is, the same representational content can be entertained by different subjects with different propositional attitudes—that one needs to represent the subject of these representational states explicitly. Thus one learns to explicitly represent oneself by learning to distinguish one's own mental and bodily states from those of others. Once the ability for explicit self-representation and the ability to self-ascribe mental and bodily states are in place, a subject can develop a "self-file." A self-file is a mental file that can be used to store all kinds of different information about oneself—including both information obtained through sources that are specific to the self, such as those based on perception, bodily experience, and introspection, and information obtained through sources that are not specific to the self, such as self-observation, reflection, and testimony. As one's self-ascriptions become increasingly differentiated and complex—ranging from basic self-ascriptions of experiences to self-ascriptions of the kinds of representations that others can have about oneself, that is to say, of meta- or even meta-metarepresentations—the self-file likewise becomes increasingly complex. The process of self-conceptualization can also be modified and enhanced by social and cultural narratives, leading to the development of a personal and cultural identity.

Thus it was shown that although theories of nonconceptual self-consciousness make an important contribution to a better understanding of the structure and development of self-consciousness, they need to be complemented by theoretical models that can make intelligible the transition from implicitly self-related information to explicit self-representation. The latter enables the ascription of mental and bodily properties to oneself as an individual entity that can be identified by others—and to which one can refer by means of the first-person pronoun to indicate this fact (cf. Tugendhat, 1979). Note that this is not to say that the subject has to identify herself to become self-conscious. Quite to the contrary, the insights that were gained through the criticism of traditional models of self-consciousness are preserved in the account presented here. Self-consciousness does not rely on self-identification; rather, we should understand it as a process of rendering explicit the self-related information that is implicit in nonconceptual representational states (in virtue of their mode of presentation), such as perception and bodily experience. This does not require self-identification, but it does require the recognition that one is a subject among others, such that one can contrast one's own states with those of others and can learn to think of oneself as an individual that can be identified and referred to by others.

Finally, chapter 7 spelled out the implications of this account for the question of whether nonhuman animals are self-aware. We saw that, to demonstrate the existence of self-awareness in nonhuman animals, what would be needed is a demonstration of the ability to explicitly distinguish between oneself and another. While much of the current research remains inconclusive (in particular due to methodological limitations inherent in nonverbal paradigms), when taken together, the existing evidence with respect to the ability for mirror self-recognition, social cognition, and metacognition in nonhuman animals suggests that at least some nonhuman animals (in particular chimpanzees and dolphins, and to a lesser extent monkeys and perhaps birds) are likely to possess abilities that can be regarded as basic forms of self-consciousness (i.e., at level 2).

## 8.2  Open Questions

We now have an account that shows how we can get from relatively primitive forms of nonconceptual representations containing implicitly self-related information to the concept of a self. However, the account also raises questions that require further research.

In particular, although chapter 5 shows how implicit information can in principle be redescribed into explicit representation, and chapter 6 illustrates this general model in more detail with regard to self-consciousness, and with the help of results from the empirical sciences, it remains an open question how exactly the process of representational redescription is supposed to work. What exactly is the mechanism of redescription, how is it initiated, and why does it occur in some cognitive systems, but not in others? Future work should aim to uncover more detail about the mechanisms involved in representational redescription, and about potential similarities and differences between humans and nonhuman animals. In doing so, the role of metacognition in particular needs to be further specified.

Further, in explaining the processes that lead to explicit self-representation, more attention needs to be paid to the relation between social cognitive abilities and self-awareness. As we saw, both the phenomenon of intersubjectivity and the relation between self-consciousness and intersubjectivity are multifaceted and rich, ranging from basic nonconceptual, procedural forms of self–other representation and differentiation to complex social phenomena and the creation of narrative and cultural identities. More research will be required to spell out these phenomena in more detail and to investigate the complex relations between them, both in humans and in nonhuman animals. To date, these abilities are still too often studied in isolation from each other, both in theoretical and in empirical work.

One of the questions that arises concerns the role of emotions for self-consciousness and social cognition. Although the emergence of self-conscious emotions and the ability for empathy seem to constitute important milestones in the development of self-consciousness and intersubjectivity, much more needs to be said about the precise role that emotions play in this development, as well as for social interaction in general. Emotions might also be involved in conferring a sense of primitive normativity to social activities, which seems to be something that human children are especially prone to (Schmidt et al., 2011; Schmidt & Tomasello, 2012).

Moreover, precisely what role language plays in these processes remains unclear. It seems that at least as far as an explicit theory of mind is concerned, linguistic abilities play a crucial role for its development. Further, as was illustrated by the example of the chimpanzee Sheba (chap. 2), the ability for symbolic, and thus more abstract, representation seems to play an important role in enabling inferential reasoning. But it is still a matter of controversy whether symbolic or linguistic abilities provide the necessary prerequisites for the development of explicit (conceptual) representations, or whether it is rather the other way around. With regard to social cognition in particular, it remains unclear whether certain social cognitive abilities enable the development of language (a position defended by Tomasello and Bogdan, e.g.), or whether it is the linguistic abilities that enable the development of complex social forms of interaction (a position defended by, e.g., Spelke). Tomasello (2008, 2009) argues that humans possess species-specific capacities and motivation for shared intentionality, which give rise to cooperative activities and linguistic communication. In this view, it is only because humans are naturally driven to share information and cooperate with each other that they are able to develop symbolic representations and linguistic communication. In other words, Tomasello sees language as the product rather than the source of unique human capacities for cooperating and communicating. In contrast, Spelke (e.g., Spelke, 2009) argues that human language is the source of unique human cognitive capacities. She points out that although infants—similar to nonhuman primates—possess an impressive array of bodies of "core knowledge," such as core representations of objects (see chap. 2) and core representations of goal-directed actions (see chap. 6), they do not begin to productively combine these representations until their second year of life. She suggests that it is the learning of words that enables young humans to join these core representations. By learning names for objects, children learn to jointly represent information that was previously represented separately, for instance, about object form and object function. The same is true, she argues, for other cognitive domains. In other words, she suggests that uniquely human forms of cognition, including forms of social cognition, depend on our capacity for combining representations productively and flexibly, and this capacity requires language. According to this view, then, it is because we have language that we can generate explicit representations, not the other way around. My own view is that both of these positions hold some truth. While we need basic social cognitive abilities to enter the realm of linguistic communication (and while these, as we have seen in chapter 6, do indeed seem to exist before children acquire a language), it is likely that more complex forms of human cognition (including more complex forms of social cognition and the development of a sophisticated self-file) depend on linguistic abilities. How exactly language and cognition influence and depend on each other is a question for further research.

Another question I have not explored in this book concerns the synchronic and diachronic integrity of the self-concept. Synchronic integrity refers to the ability to self-ascribe current mental and bodily states in a unified manner. Diachronic integrity refers to the sense of a unity of one's psychological and bodily properties over time. The latter in particular is an important prerequisite for a stable self-concept, or a sense of identity. Diachronic integrity presupposes the existence of memory, in particular episodic memory. There are interesting questions worthy of further exploration with regard to the abilities required for episodic memory and their relation to self-consciousness, and a debate is emerging about the existence of episodic memory in nonhuman animals (e.g., Griffiths, Dickinson, & Clayton, 1999). Moreover, personal narratives are also likely to play an important role in further shaping one's diachronic integrity, and although I hinted at the connection between self-consciousness and narrative practices in chapter 6, this connection was by no means fully explored. This is also true for the role that narratives play in the development of norms and group or cultural identities.

Finally, as I already pointed out in chapter 3 (sec. 3.6), the emerging field of the neuroscientific study of self-consciousness (as well as the field of social cognitive neuroscience) poses interesting questions for future research.

I hope it has become obvious that, in tackling these questions, there is both much scope and much need for interdisciplinary engagement among philosophers, psychologists, ethologists, cognitive neuroscientists, anthropologists, and others.

# Notes

## Introduction

1. Though notice that Peacocke, Recanati, and O'Brien are not explicitly concerned with developing a theory of self-consciousness based on nonconceptual content.

2. To be clear: to say that these thoughts are typically expressed by means of the first-person pronoun is not meant to exclude the possibility of self-consciousness in the absence of linguistic abilities. Whether and to what extent nonlinguistic creatures, such as animals and infants, can be self-conscious is discussed in detail in later chapters; see in particular chapters 6 and 7.

## 1 Setting the Stage: The Problem of Self-Consciousness

1. Throughout the book, I will follow the common convention of using the terms "self-awareness" and "self-consciousness" interchangeably.

2. But notice that, according to Thiel (2011), Hume is not in fact making an *ontological* claim about the nature of the self here; rather, he is making an *epistemological* claim. When he says that the mind is "nothing but a bundle or collection of different perceptions" (ibid., p. 252) and that no one can seriously claim to have "a different notion of himself" (ibid.), he is not making statements about the real nature or essence of the self. Rather, he is saying that our idea of the self must be based on what introspection or reflection reveals about it, and that the self that appears to us through introspection is that of a "bundle or collection of different perceptions" (ibid., p. 419). In Thiel's interpretation, Hume's analysis is compatible with the existence of a self beyond the perceptions, albeit a self that is inaccessible to us through inner experience.

3. It is sometimes suggested that the regress can be stopped if we assume an unconscious level of self-reflection. But this just pushes the problem to the level of the unconscious; the question regarding the source of the knowledge required to perform a judgment of identity (whether unconscious or conscious) remains. That is to

say, ultimately there must be a kind of self-awareness that does not rest on self-identification.

4. Which is not to say that all phenomenologists rejected the reflection-based account of self-consciousness. Husserl, for instance, is generally taken to have advocated a model that explains self-consciousness as a form of inner perception (e.g., Tugendhat, 1979; Frank, 1991a; Gloy, 1998); though see Zahavi (2005) for an alternative reading of Husserl.

5. See also Flanagan (1992).

6. Notice that, more recently, phenomenologically inclined authors have attempted to provide a positive account of prereflective self-consciousness in terms of bodily self-awareness (e.g., Legrand, 2007; de Vignemont, 2007). I discuss the issue of bodily (self-)awareness in detail in chapters 3 and 4.

7. For an alternative account of the "mineness" of conscious experience, see Kriegel (2009), who defends a self-representationalist theory of subjective consciousness. Note that, like the account defended here, Kriegel's account holds that the self is not explicitly represented in experience: the term "self-representational" in Kriegel's sense refers to a mental state representing itself (which he sees as a necessary condition for phenomenal consciousness), rather than to the self being part of the content of experience.

8. Thus, Tugendhat argues, pace Wittgenstein (1958) and Anscombe (1994), that although first-person statements do not self-identify, they do self-refer and do so in such a way as to imply that the speaker can, in principle, be identified by others from the third-person perspective.

9. Arguably, Tugendhat failed to see this. While he did point to an epistemic asymmetry between first- and third-person ascriptions, he did not spell out the specifics of first-person ascriptions in terms of their relevance for action, and he did not refer to a difference in state as opposed to a difference in content (Rosefeldt, 2000).

10. This example is based on an example discussed in Frank (2007).

11. According to a convention established by Castañeda (1966), the asterisk serves to indicate the s/he of self-consciousness, or the "s/he herself/himself"-locution (Chisholm, 1981).

12. This section has benefited much from discussion with Arnon Cahen.

13. Similarly, Tugendhat (1979) argues for an epistemic privilege for mental self-ascriptions only. Thus he speaks of an "epistemic asymmetry" between first- and third-personal mental-state ascriptions, in contrast to the "veritative symmetry" between them.

14. That said, it might be that only mental-state self-ascriptions are *logically* immune to error through misidentification. The distinction between logical and de facto immunity and its significance is discussed in chapter 4.

15. Compare this to the famous example of Mach, who once entered a bus and, upon seeing himself in the bus's rear mirror, thought, "What a shabby pedagogue," without realizing that the man he saw in the mirror was himself.

16. In addition to the distinction between ownership and agency, Synofzik, Vosgerau, and Newen (2008) also suggest distinguishing between a (nonconceptual) *sense* or *feeling* of ownership/agency and a (conceptual) *judgment* of ownership/agency. We can ignore these details for the purposes of this chapter.

17. Other authors have suggested that what is lacking in cases of thought insertion is not the sense of agency, but rather an endorsement of or commitment to the thought in question (Bortolotti, 2010; Fernández, 2010).

18. This is not to say that the phenomenon of thought insertion might not have extremely interesting philosophical implications. In particular, it might have important implications for theories that are concerned with the phenomena of rational agency and authority over one's thoughts (e.g., Moran, 2001).

19. I owe the following argument to Arnon Cahen.

20. One might reject the analyses that are based on this distinction because, for example, one might find it questionable whether thoughts should be modeled according to models for motor action at all.

21. Coliva (2002) makes a similar point. She argues that reports of thought insertion show at best that it is conceivable to make third-person ascriptions on the basis of introspective awareness of a mental state. But this does not imply that introspection-based *self*-ascriptions of mental states are not immune to misidentification.

22. I take it that similar considerations will apply to the phenomenon of anarchic hand syndrome.

23. Henrich lodged a similar criticism against Tugendhat's analysis of self-consciousness in terms of the ability to self-refer by means of the first-person pronoun. According to Henrich, the ability to use the first-person pronoun already presupposes consciousness of oneself as an identifiable entity in the world (Rosefeldt, 2000).

24. See Hamilton (2013) for a recent account that argues for such a "conceptual holism" between self-consciousness and linguistic self-reference.

## 2 Nonconceptual Content

1. Around the same time, Dretske (1981) also introduced the distinction between analog and digital representation. I return to this distinction in the following section. It is important to note, though, that in Evans's view, informational states have nonconceptual content before becoming conscious, and they become conscious

only when they become conceptualized. Thus his notion of nonconceptual content differs importantly from the notion defended here—and from the way the notion is discussed in much of the literature.

2. Of course, some scholars deny this and are willing to ascribe concepts to beings already on the basis of basic discriminatory abilities. However, in such a wide notion of concept possession, the notion of concepts becomes virtually useless for answering questions about the specific nature of human thought.

3. Note that the notion of concept possession admits of degrees. That is to say, there might be cases in which a subject has a partial, but not full, grasp of a concept. This would be evidenced, for example, by the fact that they possess some, but not all, of the abilities mentioned.

4. Conceptualists like McDowell (1994) and Brewer (1999) argue for conceptualism on the grounds that (1) perception can rationally justify beliefs, and (2) only states with conceptual content can rationally justify beliefs. For a detailed rebuttal of this view, see Hopp (2011). In contrast to other authors who have defended nonconceptualism against this charge (e.g., Peacocke, 1992; Heck, 2000), Hopp makes the opposite case: in his view, we can only understand how perception can justify belief if we accept that it has nonconceptual content.

5. This is related to the idea that the content of perceptual states is analog in nature, whereas the content of propositional attitudes is digital (Dretske, 1981).

6. Some authors have recently argued that to explain how experience can make conceptual (i.e., general, abstract) thought possible, we need to assume a relational view of experience (e.g., Campbell, 2002; Brewer, 2006). In this view, perception does not have representational content at all. Rather, experience of an object consists in standing in a certain relation to the object, which makes the object available to thought but does not entail any representation of it. For forceful rebuttals of this view, see Ginsborg (2011a) and Hopp (2011). Alternatively, one might attempt to give an account of how the concept that is being elucidated can be built out of other, more primitive concepts. However, while this strategy might work for some concepts, it is implausible to assume that it will work for all concepts. See Roskies (2008) for an argument about why this strategy fails in the case of perceptual concepts, for example.

7. Note, though, that the term "normativity" as it is used in this context is somewhat stronger than the way it was used in the context of sorting red objects. Here the term appeals to primitive normativity in the sense of a primitive rationality, or "reasons"-responsiveness, which is stronger than the notion of normativity in the sense of the correct or incorrect use of a term (see Ginsborg, 2011b). I take both types of primitive normativity to be a case in point of nonconceptualism.

8. This is related to a proposal recently made by Hopp (2011), according to which nonconceptual content is defined by nondetachability from its present context. In contrast, conceptual content is detachable from the present context.

9. The interpretation that the behavioral differences in these contexts is due to an executive function impairment (i.e., the inability to inhibit food-approaching behavior when faced with the candies) seems to be ruled out by a follow-up experiment. This experiment (Boysen, Mukobi, & Berntson, 1999) used mixed candy/numeral choice pairs in addition to candy/candy and numeral/numeral pairs. Chimps were as successful with mixed candy/numeral pairs as with numeral/numeral pairs. If the inhibition interpretation was correct, one would expect chimps to show a response bias toward the candy in the mixed pairs, but this was not found (see Hurley, 2006, p. 159).

10. According to the *cohesion principle*, two surface points lie on the same object only if the points are linked by a path of connected surface points. This implies that all points on an object move on connected paths over space and time. According to the *boundedness principle*, two surface points lie on distinct objects only if no path of connected surface points links them. That is to say, when surfaces are neither spatially separated nor separately moving, they are part of the same object. According to the *principle of rigidity*, two surfaces that undergo distinct rigid motions are perceived as distinct bodies. Finally, according to the principle of *no action at a distance*, two surfaces that undergo a common rigid motion are perceived as connected, unless there is a detectable separation between them (Spelke, 1990).

11. Alternatively, a proponent of the state view could hold that all content is nonconceptual. Such a position seems to be defended by Stalnaker (1998), who takes all content to be informational (possible worlds) content. This is not the way in which the issue is generally discussed in the debate about nonconceptualism, however, so this view can be neglected for the purposes of this chapter.

12. An alternative would be to characterize nonconceptual content in terms of Peacocke's (1992) "positioned scenario content." This consists essentially in a systematic specification of the ways in which a space can be filled out around a perceiver. However, this notion does not seem to do proper justice to the subject's apprehension of her possibilities of interacting with the environment. Moreover, it seems unable to account for the partial indeterminacy of visual perception (Noë, 2002), and it is unclear how it could be extended beyond the domain of visual perception (Bermúdez & Cahen, 2008).

13. Stanley and Williamson (2001) have put forward a much-discussed argument against the thesis that a fundamental distinction exists between knowledge-how and knowledge-that. They claim that knowledge-how is simply a species of knowledge-that. But see, e.g., Noë (2005) and Rosefeldt (2004) for convincing rebuttals of this claim.

14. Notice, however, that Milner and Goodale (1995) show that fast grasping behavior conforms to the actual length of the lines. So there is a sense in which we implicitly take ourselves to be able to perform actions that imply a different length of the two lines (as reflected in our visual awareness of the two lines), and another sense in which we implicitly take ourselves to be able to perform actions that imply the same length of the two lines (as reflected in our reaching behavior).

## 3   Self-Representationalist Accounts of Nonconceptual Self-Consciousness

1. Though notice that Gibson himself defends an antirepresentationalist theory.

2. See also Schwenkler (2014) for a recent defense of this view.

3. In the rubber hand illusion, subjects are placed in front of a table facing a life-size rubber model of a hand, while their own hand is hidden from view by a screen. When both the rubber hand and the real hand are simultaneously stroked by a brush, subjects report experiencing the sensation in the location of the rubber hand, rather than their real hand (Botvinick & Cohen, 1998).

4. In addition to agent-relative knowledge and self-attached knowledge, Perry also distinguishes knowledge of the person one happens to be. This is a kind of knowledge that is in fact about oneself but is not recognized as such by the thinker. We have encountered examples of this kind of knowledge in chapter 1, for instance, when we discussed the possibility of seeing oneself in the mirror without realizing that the person one sees is oneself.

5. See also Cassam (1994, p. 52), who argues that "in egocentric spatial perception the objects of perception are experienced as standing in spatial relations *to the perceiver.*"

6. It does not matter for this argument whether squirrels are in fact able to distinguish their perspective from that of others or not. If you happen to think that squirrels are able to do so, just insert an organism of your choice that you think is not able to distinguish its own perspective from that of others. I will return to the issue of "perspectival differentiation" in later chapters. See also chapter 7 for discussion of the question of whether and to what extent nonhuman animals possess self-awareness.

7. See also Millikan (1990) for a related point.

8. The relevance of Perry's considerations to the issues at hand was first brought to my attention by a conference presentation given by Arnon Cahen (2006). It is also briefly discussed (and dismissed) by Meeks (2006). See also Recanati (2007).

9. Note that in his essay "Thought without Representation," Perry states that "Z-land comes in not as an unarticulated constituent each Z-landish weather state-

ment is about, but as a global factor that all Z-land discourse about the weather concerns" (Perry, 2000, p. 179). This is to distinguish the Z-land case from cases where we talk about weather without articulating a location because it is obvious from the conversational context what the location is; in our case, the location does come in as an unarticulated constituent of each weather statement. In contrast to Z-landers, we are aware of other places and thus have to "track" which location the conversation is about (whether or not that location is articulated). But since Z-landers are not aware of other places, no such tracking has to take place for them; their beliefs and assertions carry a "lesser burden" compared to ours, so to speak. Accordingly, "the Z-landers' assertions and beliefs concern Z-land, but are not about Z-land" (Perry, 2000, p. 179). What I am claiming here is that the case of perception and bodily experience is analogous to the case of weather reports in Z-land.

10. Schear (2009) makes a similar point by providing a discussion of Sartre's phenomenological analysis of the experience of being immersed in reading a book.

11. Awareness of a fact or state of affairs requires that this fact or state of affairs be explicitly represented. I discuss the relation between awareness and explicit representation in detail in chapter 5.

12. Notice that this argument is actually independent of the question of whether the content of perception and proprioception is indeed best characterized as being nonconceptual. Although, as I argued earlier, we have good reasons to think that this is the case, even in a conceptualist theory of perception and proprioception, the same point still applies. Whether or not the content of perception and proprioception is conceptual or nonconceptual, the fact remains that no explicit self-representation is required to account for the action-guiding function of this content.

13. Schwenkler (2014) has recently argued that the use of monadic egocentric notions is insufficient to characterize the content of visual kinesthesis. In other words, he seems to suggest that the self is part of the content of visual perception. However, I am unsure of how strong this claim is. Indeed, it seems to me that the perspectival or self-locating nature of visual experience to which Schwenkler appeals to explain self-motion can be accounted for in the view defended here, namely, by taking into account the role of mode of perception. So the suggestion is that the experience of self-motion is just the experience of motion in a particular way, that is, via a self-specific or internal mode of presentation. This mode of presentation accounts for the particular character of the experience. However, this does not require that the self is explicitly represented in the content of experience. I discuss the role of the mode of experience in more detail in the next chapter.

14. I owe thanks to Arnon Cahen for helping me appreciate this point.

15. Or, as we shall see in chapter 5, at least in the realm of explicit object-predicate presentation.

## 4   Toward a Non-Self-Representationalist Account

1. Boyle (2011) seems to make a related point when he claims that a subject who possesses psychological concepts "merely knows how to make explicit aspects of what she is already presupposing" (p. 10).

2. Notice, however, that Peacocke restricts this claim to particular mental states only.

3. See also Coliva (2012) in the same volume for a related discussion; the same considerations would apply to her view (as well as to the arguments put forward by Schwenkler [2014]; see the previous chapter).

4. Note that it is perfectly possible to represent oneself as an intentional object—as is the case when one looks at oneself in a mirror, for example. In such a case, the self plays a double role: once as the subject of perception and a second time as the object of perception. In the latter role, it can be misrepresented, while in the former it cannot.

5. I am grateful to Glenn Carruthers for helpful discussions of this point.

6. In contrast to this analysis, Schear (2009) seems to put Zahavi in the self-representationalist camp and on this ground argues against his view. I think that it is indeed not always clear whether Zahavi and other phenomenologists see the sense of "mineness" as something above and beyond the content of experience in combination with its mode (as Schear's analysis seems to suggest), or whether they would be happy with the deflationary account I have presented here. But it seems to me that it is at least not impossible in principle to reconcile their views with an analysis that is based on non-self-representationalism.

7. Note that this should not lead us to the conclusion that the self does not exist—although Hume is often read in this sense.

8. Hence one might say that one aspect of the pathology in patients with schizophrenia who are suffering from thought insertion is that their thinking no longer complies with this logical point.

9. Though see Hamilton (2013) for a critical discussion of arguments based on quasi proprioception.

10. According to O'Brien, the problem with a self-representationalist (or content-based) approach is that it cannot account for the possibility of self-reference in the absence of self-specifying information provided by the senses, as in Anscombe's (1975) sensory deprivation tank scenario.

11. For related accounts that locate the notion of rational agency and agent's awareness at the heart of the phenomenon of self-consciousness, see Moran (2001) and Rödl (2007).

12. Accordingly, in O'Brien's view, agent's awareness is a case in point for the "traditional view" that "we have self-knowledge only of our psychological properties." However, because she takes actions to be physical events, such knowledge "can also deliver knowledge of our physical properties" (O'Brien, 2007, p. 127).

13. Of course, one might say that insofar as we can only ascribe representational content to intentional agents, the notion of intentional agency is fundamental to all these propositional attitudes. Moreover, there are other ways in which the fact that we are agents feeds into the determination of the content of our representational states (see chap. 2). In other words, in some sense representation—including self-representation—can only be understood in relation to intentional agency. However, I take it that this is not the point O'Brien is trying to make. After all, her aim is precisely to single out and distinguish the notion of agent's awareness from other forms of awareness, such as bodily experience.

14. Moreover, as Hamilton (2013) points out, it is not clear what our intuitions should be in response to such cases.

15. See also Legrand (2006) for an account of bodily awareness as awareness of oneself "as subject."

16. Note that I do not mean to suggest that there might not also be important differences between awareness of ourselves as agents and awareness of ourselves as patients. Indeed, as Boyle (2009, 2011) suggests, we might have good reasons to distinguish spontaneous from receptive self-knowledge. Still, both are important forms of self-consciousness.

17. These include the temptation to identify the self with a mental entity, as Descartes did, as well as the temptation to identify "the self with a sequence of appropriately related mental states, as neo-Lockeans do nowadays; or else, to embrace either the view that there is no bearer of such mental states, as Hume did; or to deny, with Wittgenstein and Anscombe, that 'I' is a genuinely referring expression" (Coliva, 2012, p. 25). (See also chap. 1 for a discussion of some of these "illusions of transcendence.")

18. In a sense, this echoes Evans's (1982) contention that the recognition that bodily self-ascriptions can be immune to error through misidentification provides the best "antidote" to Cartesian dualism.

## 5   From Implicit Information to Explicit Representation

1. I am grateful to Ulman Lindenberger for drawing my attention to the relevance of Karmiloff-Smith's theory for my project.

2. Kirsh also suggests that this immediacy can be measured in terms of computational processing. He proposes that the degree to which a given representation R

explicitly encodes information I, for a given creature C, should be measured by the amount of computation C must perform to extract I (Kirsh, 1990, 2009).

3. The model thus attempts to reconcile a (moderate) form of nativism and domain specificity with a domain-general perspective, such as the one advocated by Piaget.

4. Notice, however, that human rationality also often appears to be context bound (Hurley, 2006). Consider, for instance, Cosmides and Tooby's (1992) work on people's abilities to apply "if p then q" rules for reasoning. While people are successful at this task when asked to detect "cheaters," they fail to correctly apply the rule in other contexts.

5. Further empirical support is provided by the numerous other experiments described in Karmiloff-Smith's book and elsewhere. Notice that the age-related stages described in this experiment will be different when it comes to other cognitive abilities.

6. Note that empirically, it is not always possible to distinguish between levels E2 and E3, which is why they are generally taken together as E2/E3.

7. Notice also that in some cases, the process of representational redescription may take the opposite direction. For instance, a person who is just learning to drive a car receives explicit instructions from her driving instructor and thus will start off with explicit representations that guide the way she behaves behind the steering wheel. As the new driver acquires more skill, these explicit representations will then bit by bit be transformed into a format in which the information is encoded in procedures and thus remains implicit in the activity. This will enable more fluid driving but will at the same time prevent the information from being consciously accessible. (See Baars [1988] for a theory of how this process occurs. Another example of this process would be learning to touch-type.) Thus we find two complementary directions in learning and development. "On the one hand, they involve the gradual process of proceduralization (that is, rendering behavior more automatic and less accessible). On the other hand, they involve a process of 'explicitation' and increasing accessibility (that is, representing explicitly information that is implicit in the procedural representations sustaining the structure of behavior)" (Karmiloff-Smith, 1996, p. 17). So the process of representational redescription in general is bidirectional, although for our specific purposes, the direction from implicit to explicit is the more interesting one.

8. For a computational model of the transition from implicit to explicit representations, see Sun et al. (2001).

9. For the difference between representing an object and its property and representing a predication relation between object and property, see Bogdan (2009).

10. More generally, we can speak of intentional relations and can distinguish between action-oriented intentional relations, emotional intentional relations, and

epistemic intentional relations. See chapter 6 for more details on these distinctions. For the sake of simplicity, I will here follow the standard terminology of propositional attitudes, although, as mentioned in the introduction, I generally prefer to speak of intentional relations.

## 6   Self and Others, or The Emergence of Self-Consciousness

1. Nonetheless, for readers wanting to use the term in a stricter sense, I will indicate from what point onward in the developmental trajectory the term may be used in this stricter sense (see sec. 6.3.3).

2. Naturally, it is important to keep in mind that the different historical proposals regarding intersubjectivity have to be understood in the respective contexts in which they were being developed and the systematic roles they played for the philosophical frameworks of the respective authors. For instance, while some authors pursued predominantly epistemological or metaphysical questions, others took an ethical, political, or linguistic interest in the phenomenon of intersubjectivity. Naturally, these different perspectives have implications for the methodology and the philosophical focus applied by different authors. These substantial differences need not concern us here, but it would obviously be important for a systematic analysis of historical approaches to bear them in mind.

3. Fichte's view, in turn, goes back to Kant (2009). According to Kant, self-consciousness is the condition of the possibility of the activities of reason. Accordingly, as the foundation of object cognition, the self cannot grasp itself as an object (and thus cannot reflect on itself) without thereby presupposing itself. This insight was subsequently developed by Fichte, with whom self-consciousness also entered into the focus of philosophical theorizing for its own sake, rather than for the sake of the role self-consciousness played in the epistemological projects that were central for Descartes and Kant.

4. For a detailed discussion of Fichte's position, see also, e.g., Crone (2005).

5. If this is right, then the notion of rational agency indeed lies at the heart of self-consciousness (see Moran, 2001; O'Brien, 2007; and Rödl, 2007). However, this also implies that we cannot explain rational agency without intersubjectivity. Moreover, I would argue that the ability to recognize the other as a subject who is "addressing" me as a rational agent presupposes more basic abilities of self–other differentiation, which also need to be accounted for.

6. It is interesting to note that Sartre emphasizes the affective component of intersubjective encounters, in particular the feelings of shame and guilt. We will briefly return to these emotions and their significance for self-consciousness later. For a more detailed discussion, see Rochat (2009).

7. This view, in turn, seems to go back to insights developed by Husserl (Zahavi, 2005).

8. According to Habermas (1999), Hegel was the first to introduce language, work, and symbolic interaction as the media through which the human mind is formed and transformed, and, according to Habermas, this was subsequently picked up and further developed by thinkers such as Humboldt, Peirce, Dilthey, Dewey, Cassirer, Heidegger, Wittgenstein, and Davidson (though the claim is not that these thinkers were all directly influenced by Hegel).

9. For a detailed discussion of Mead's account of the relation between self-consciousness and intersubjectivity, see Lailach-Hennrich (2011).

10. While Strawson restricts this claim to mental-state concepts, in my view, the same is true of our concepts of bodily states.

11. Bermúdez (1998) argues against this view. Instead he proposes a different argument for the constitutive relation between self-consciousness and intersubjectivity. For a critical discussion, see Dow (2012).

12. Moreover, if we take seriously Wittgenstein's argument against the possibility of a private language, one cannot possibly possess a concept that is only being applied to oneself, in the absence of external (i.e., intersubjective) criteria for its application (see Wittgenstein, 1967).

13. Notice also that Bermúdez's claim is limited to psychological self-ascriptions. As was shown in previous chapters, he does not believe that the ability for social cognition is required for the self-ascription of bodily states.

14. See also Dow (2012) and Lailach-Hennrich (2011) for arguments against Bermúdez's account of joint engagement.

15. Bogdan (2010) has also recently proposed a theory that links self-consciousness to social cognitive abilities. However, Bogdan's focus and approach are different from the focus and approach of this book. For one thing, his target seems to be consciousness in general, rather than self-consciousness in particular. (Indeed, he does not seem to see a distinction between consciousness and self-consciousness.) Moreover, his approach is to uncover the distinct evolutionary pressures that have led to the emergence of human consciousness.

16. One might argue that insofar as we are dealing with information about other *subjects* (rather than objects), it would be more appropriate to talk about a second-person perspective. For a detailed discussion of the notion of a second-person perspective, see, e.g., Rödl (2007, chap. 6) and Pauen (2012). While this is an important point to make, for the sake of simplicity, I will ignore this rather complex issue here and only refer to the distinction between first- and third-person perspective.

17. Notice how similar ideas were already expressed by Merleau-Ponty (1962), who, as we saw earlier, argued for the existence of a prereflective recognition of one's

body as being similar to that of others, which he attributed to a common "corporeal schema."

18. Similar results, albeit of a highly transient nature, were recently found in newborn monkeys (Ferrari et al., 2006).

19. See Heyes (2009) for an account that suggests that mirror neurons develop on the basis of sensorimotor learning processes, rather than constituting a specialized and innate module for imitation and action recognition.

20. That said, Rochat and Hespos (1997) presented evidence that newborn infants show differential rooting responses to self-generated touch compared with being touched by another. This seems to suggest an inborn ability to distinguish between self and other when it comes to bodily experience. However, these responses could also be based on the detection of multimodal contingencies (similar to the results found in the study by Bahrick and Watson) without necessarily implying a representation of self as distinct from other. Nonetheless, both the results found by Rochat and Hespos and those found by Bahrick and Watson could point toward mechanisms that provide the basis for bodily self-awareness, and it is possible that bodily self-awareness emerges before psychological self-awareness.

21. In fact, Tomasello argues that it is precisely the ability and desire for cooperation, enabling the development of "we-intentions" (Tuomela & Miller, 1988), that distinguish humans from other primates and account for language learning and other cultural achievements (e.g., Tomasello, 2009).

22. Recall that this affective component of intersubjectivity and self-consciousness was already emphasized by Sartre.

23. Notice that some authors distinguish between simple and cognitive empathy, rather than between emotional contagion and empathy (e.g., Preston & de Waal, 2003). According to them, cognitive empathy implies a differentiation between self and other (subject and object, in their terminology), while simple empathy does not require such a differentiation. Thus the term "empathy" as I use it here is equivalent to the term "cognitive empathy" in that alternative terminology.

24. Children at this age may even begin to show an understanding that others can have "false beliefs," that is, beliefs that misrepresent a state of affairs and that differ from what the child knows to be the case. For instance, a study by Buttelmann, Carpenter, and Tomasello (2009) has shown that eighteen-month-old infants in an active helping paradigm were able to take into account an adult's false belief in order to determine his goal. This is surprising, as false belief understanding is traditionally held to emerge only at the age of four to five years (see sec. 6.3.4). However, it is unclear whether the false-belief understanding displayed by eighteen month olds is the same as that displayed by older children (see sec. 6.3.4 for further discussion).

25. See the previous chapter for a discussion of the notion of a "theory-in-action."

26. For a detailed discussion of the notion of "encountering," see Apperly and Butterfill (2009) and Butterfill and Apperly (2013).

27. Notice also that the conceptual understanding of beliefs and other mental states continues to develop at later ages. For instance, five- and six-year-old children still struggle to understand how beliefs are acquired (Carpendale & Chandler, 1996; Robinson & Apperly, 2001) or how they interact with other mental states, such as desires (Leslie et al., 2005; Leslie & Polizzi, 1998).

28. Moreover, insofar as simulation theory presupposes the ability to ascribe mental states to oneself before this ability can be applied to others, it is incompatible with the constitutive relation between self-consciousness and intersubjectivity that I am suggesting here.

29. For a discussion of the relation between simulation theory and mirror neurons, see Gallese and Goldman (1998) and Barresi and Moore (2008).

30. Whether this interpretation can withstand scrutiny will depend on what kind of version of theory-theory and narrative theory one adopts. Various versions of each theory are in play in the literature (in particular theory-theory), which renders any evaluation of their compatibility problematic, for not all versions will fit into the general framework I am proposing. Nonetheless, generally speaking, insofar as it is reasonable to say that there are certain theoretical principles that people employ to make sense of other people's behavior (e.g., in terms of belief-desire explanations), the existence of such principles warrants the talk of a theory of mind in the theory-theory sense. Likewise, insofar as the way we make sense of people's behavior is importantly structured by the kind of narratives we tell about them, there is theoretical room for a narrative practice approach to mindreading. However, (a) it is not clear why we should always need to refer to a narrative to make sense of someone's behavior (and it may not always be possible to actually come up with a suitable narrative); and (b) insofar as social narratives are supposed to confer basic principles of human behavior that can subsequently be used in a variety of situations, the narrative approach seems to be compatible with a theory-theory approach to mindreading as it is being understood here. Thus it seems to me a reasonable view that both approaches are compatible with (in the sense of being complementary), rather than mutually exclusive of, each other (and of simulation and interaction theory), even if the narrative theorist would disagree (see Hutto, 2008).

31. Some authors have recently suggested another alternative to standard theories of social cognition, namely, so-called person model theories or person perception theories (e.g., Newen & Schlicht, 2009; Dow, 2012; Barresi, Moore, & Martin, 2013; see also Andrews, 2012). Although these are generally construed as alternatives to the standard theories (while I have argued for an integrative approach), I take them to be largely compatible with the view defended here, with respect to an acknowledgment both of the constitutive link between self-consciousness and intersubjectivity (in the spirit of the Strawsonian view) and of the multifaceted nature of social

cognition. See also Fiebich and Coltheart (in press) for another recent pluralistic approach to social cognition. Notice also that pluralistic approaches, such as the one defended here, go beyond the two-systems approach mindreading that has been proposed by Apperly and Butterfill (2009). It is an interesting question for future research whether and how pluralistic approaches can be reconciled with a two-systems approach.

32. Andrews also argues that the attribution of mental states to others plays much more of a role in explaining behavior than in predicting behavior. This is certainly an interesting suggestion, which future research should explore further.

33. See Hurlburt and Schwitzgebel (2007) for a detailed description and critical discussion of the method.

34. In this task, participants are asked to fetch a Band-Aid for the experimenter (who pretends to have cut her finger) from a selection of three boxes, one of which is labeled as a Band-Aid box. Upon opening the box, the participant finds out that the Band-Aid box actually contains candles. The participant is then asked a self-question ("Before you looked inside the box, what did you think was inside?"), an other-question ("Later I will show this box to your teacher. He hasn't seen inside here, though. What will he think is in there?"), and a reality control question ("What's in the box, really?"). The result of this study is fascinating and surprising because it suggests that the ability to ascribe false beliefs to others might be more basic than the ability to ascribe them to oneself. However, it is not clear how to interpret this result, and it remains to be seen whether it can be generalized, as the acquisition of theory of mind in individuals with autism spectrum disorder might follow a developmental trajectory that could be quite different from that followed by typically developing children. For an account of why ascribing mental states to oneself might be more demanding than ascribing them to others, see Bogdan (2010).

## 7   Self-Consciousness in Nonhuman Animals

1. Note that these results are yet to be independently replicated (cf. Suddendorf & Butler, 2013).

2. For a detailed discussion of the problem of distinguishing between mental-state attributions and behavior reading, see Lurz (2011).

3. This is not to say that humans never perform genuinely higher-order judgments, of course.

4. If this is right, studies involving metacognitive tasks are actually not very well suited for studying self-consciousness, as they do not assess direct access to one's own mental states.

5. Similar considerations apply to experiments that test information-seeking behavior in cases of uncertainty.

6. But notice that, as we saw in the previous chapter, nonverbal false-belief tasks are difficult to interpret and do not necessarily imply a conceptual understanding of beliefs. For one thing, they are generally unable to rule out competing behavior-reading explanations. For another, they can be explained based on the ascription of mindreading abilities that do not amount to a full-fledged grasp of mental-state concepts.

7. Of course, the same is true for nonverbal paradigms used to study social cognitive abilities in infants, as we saw in the previous chapter.

# References

Amsterdam, B. (1972). Mirror self-image reactions before the age of two. *Developmental Psychobiology*, *5*, 297–305.

Anderson, J. R., & Gallup, G. G., Jr. (1999). Self-recognition in non-human primates: past and future challenges. In M. Haug & R. E. Whalen (Eds.), *Animal models of human emotion and cognition* (pp. 175–194). Washington, DC: American Psychological Association.

Anderson, M., & Perlis, D. (2005). The roots of self-awareness. *Phenomenology and the Cognitive Sciences*, *4*, 297–333.

Andrews, K. (2012). *Do apes read minds?* Cambridge, MA: MIT Press.

Anscombe, G. E. (1994). The first person. In Q. Cassam (Ed.), *Self-knowledge* (pp. 140–159). Oxford: Oxford University Press. Original work published in S. Guttenplan (Ed.), *Mind and language: Wolfson College Lectures* (pp. 45–65). Oxford: Oxford University Press, 1975.

Apperly, I., & Butterfill, S. (2009). Do humans have two systems to track beliefs and belief-like states? *Psychological Review*, *116*(4), 953–970.

Armel, K. C., & Ramachandran, V. S. (2003). Projecting sensations to external objects: Evidence from skin conductance response. *Proceedings of the Royal Society of London. Series B, Biological Sciences*, *270*(1523), 1499–1506.

Astington, J. W., & Jenkins, J. M. (1999). A longitudinal study of the relation between language and theory-of-mind development. *Developmental Psychology*, *35*(5), 1311–1320.

Baars, B. (1988). *A cognitive theory of consciousness*. New York: Cambridge University Press.

Bahrick, L. E., & Watson, J. S. (1985). Detection of intermodal proprioceptive-visual contingency as a potential basis of self-perception in infancy. *Developmental Psychology*, *21*(6), 963–973.

Baillargeon, R. (1987). Young infants' reasoning about the physical and spatial properties of a hidden object. *Cognitive Development, 2*(3), 179–200.

Baker, L. R. (1998). The first-person perspective: A test for naturalism. *American Philosophical Quarterly, 35*(4), 327–348.

Baker, L. R. (2012). From consciousness to self-consciousness. *Grazer Philosophische Studien, 84*, 19–38.

Baron-Cohen, S. (1995). *Mindblindness*. Cambridge, MA: MIT Press.

Baron-Cohen, S., Leslie, A. M., & Frith, U. (1985). Does the autistic child have a theory of mind? *Cognition, 21*(1), 37–46.

Barresi, J., & Moore, C. (1996). Intentional relations and social understanding. *Behavioral and Brain Sciences, 19*(1), 107–122.

Barresi, J., & Moore, C. (2008). The neuroscience of social understanding. In J. Zlatev, T. Racine, C. Sinha, & E. Itkomen (Eds.), *The shared mind: Perspectives on intersubjectivity* (pp. 39–66). Amsterdam: John Benjamins.

Barresi, J., Moore, C., & Martin, R. (2013). Conceiving of self and others as persons: Evolution and development. In J. Martin & M. Bickhard (Eds.), *The psychology of personhood: Philosophical, historical, social-developmental, and narrative perspectives* (pp. 127–146). Cambridge: Cambridge University Press.

Bates, E. (1979). *The emergence of symbols: Cognition and communication in infancy*. Waltham: Academic Press.

Beckermann, A. (2003). Self-consciousness in cognitive systems. In C. Kanzian, J. Quitterer, & E. Runggaldier (Eds.), *Persons: An interdisciplinary approach* (pp. 72–86). Wien: ÖBV-hpt.

Bermúdez, J. (1995). Nonconceptual content: From perceptual experience to subpersonal computational states. *Mind & Language, 10*, 333–369.

Bermúdez, J. (1998). *The paradox of self-consciousness*. Cambridge, MA: MIT Press.

Bermúdez, J. (2003). *Thinking without words*. Oxford: Oxford University Press.

Bermúdez, J. (2007). What is at stake in the debate on nonconceptual content? *Philosophical Perspectives, 21*(1), 55–72.

Bermúdez, J., & Cahen, A. (2008). Nonconceptual mental content. In Edward N. Zalta (Ed.), *The Stanford encyclopedia of philosophy* (Spring 2008 Ed.), http://plato.stanford.edu/archives/spr2008/entries/content-nonconceptual.

Bischof-Köhler, D. (1988). Über den Zusammenhang von Empathie und der Fähigkeit, sich im Spiegel zu erkennen. *Schweizerische Zeitschrift für Psychologie, 47*, 147–159.

Bogdan, R. J. (2009). *Predicative minds: The social ontogeny of propositional thinking.* Cambridge, MA: MIT Press.

Bogdan, R. J. (2010). *Our own mind: Sociocultural grounds for self-consciousness.* Cambridge, MA: MIT Press.

Bortolotti, L. (2010). *Delusions and other irrational beliefs.* Oxford: Oxford University Press.

Botvinick, M., & Cohen, J. (1998). Rubber hands "feel" touch that eyes see. *Nature, 391*(6669), 756.

Boyle, M. (2009). Two kinds of self-knowledge. *Philosophy and Phenomenological Research, 78*(1), 133–164.

Boyle, M. (2011). Transparent self-knowledge. *Aristotelian Society, Supplementary Volume, 85*(1), 223–241.

Boysen, S. T., & Berntson, G. G. (1995). Responses to quantity: Perceptual versus cognitive mechanisms in chimpanzees (Pan troglodytes). *Journal of Experimental Psychology: Animal Behavior Processes, 21*(1), 82–86.

Boysen, S. T., Mukobi, K. L., & Berntson, G. G. (1999). Overcoming response bias using symbolic representations of number by chimpanzees (Pan troglodytes). *Animal Learning & Behavior, 27*(2), 229–235.

Brewer, B. (1995). Bodily awareness and the self. In N. Eilan & J.-L. Bermúdez (Eds.), *The body and the self* (pp. 291–303). Cambridge, MA: MIT Press.

Brewer, B. (1999). *Perception and reason.* New York: Oxford University Press.

Brewer, B. (2006). Perception and content. *European Journal of Philosophy, 14,* 165–181.

Buttelmann, D., Carpenter, M., & Tomasello, M. (2009). Eighteen-month-old infants show false belief understanding in an active helping paradigm. *Cognition, 112*(2), 337–342.

Butterfill, S., & Apperly, I. (2013). How to construct a minimal theory of mind. *Mind & Language, 28*(5), 606–637.

Byrne, A. (2003). Consciousness and nonconceptual content. *Philosophical Studies, 113*(3), 261–274.

Cahen, A. (2006). The implicit self in perception. Poster, ASSC10 conference, Oxford, UK.

Call, J., & Tomasello, M. (1999). A nonverbal false belief task: The performance of children and great apes. *Child Development, 70*(2), 381–395.

Call, J., & Tomasello, M. (2008). Does the chimpanzee have a theory of mind? Thirty years later. *Trends in Cognitive Sciences, 12*(5), 187–192.

Camp, E. (2009). Putting thoughts to work: Concepts, systematicity, and stimulus-independence. *Philosophy and Phenomenological Research, 78*(2), 275–311.

Campbell, J. (1994). *Past, space, and self.* Cambridge, MA: MIT Press.

Campbell, J. (1999). Schizophrenia, the space of reasons, and thinking as a motor process. *Monist, 82*(4), 609–625.

Campbell, J. (2002). *Reference and consciousness.* Oxford: Oxford University Press.

Carey, S. (2009). *The origin of concepts.* New York: Oxford University Press.

Carpendale, J. I., & Chandler, M. (1996). On the distinction between false belief understanding and subscribing to an interpretive theory of mind. *Child Development, 67*, 1686–1706.

Carpenter, M., Akhtar, N., & Tomasello, M. (1998). Fourteen- through 18-month-old infants differentially imitate intentional and accidental actions. *Infant Behavior and Development, 21*(2), 315–330.

Carruthers, G. (2008). Commentary on Synofzik, Vosgerau and Newen (2008). *Consciousness and Cognition, 18*, 515–520.

Carruthers, P. (2008). Meta-cognition in animals: A skeptical look. *Mind & Language, 23*, 58–89.

Carruthers, P. (2011). *The opacity of mind: An integrative theory of self-knowledge.* Oxford: Oxford University Press.

Cassam, Q. (1994). *Self and world.* New York: Oxford University Press.

Castañeda, H. N. (1966). He: A study in the logic of self-consciousness. *Ratio, 8*(2), 130–157.

Chisholm, R. M. (1981). *The first person: An essay on reference and intentionality.* Minneapolis: University of Minnesota Press.

Christoff, K., Cosmelli, D., Legrand, D., & Thompson, E. (2011). Clarifying the self: Response to Northoff. *Trends in Cognitive Sciences, 15*(5), 187–188.

Chuard, P. (2007). The riches of experience. *Journal of Consciousness Studies, 14*(9–10), 20–42.

Clayton, N. S., Dally, J. M., & Emery, N. J. (2007). Social cognition by food-caching corvids: The western scrub-jay as a natural psychologist. *Philosophical Transactions of the Royal Society of London: Series B, Biological Sciences, 362*(1480), 507–522.

Coliva, A. (2002). Thought insertion and immunity to error through misidentification. *Philosophy, Psychiatry, & Psychology, 9*(1), 27–34.

Coliva, A. (2003). The argument from the finer-grained content of colour experiences: A redefinition of its role within the debate between McDowell and non-conceptual theorists. *Dialectica*, *57*(1), 57–70.

Coliva, A. (2006). Error through misidentification: Some varieties. *Journal of Philosophy*, *103*(8), 403–425.

Coliva, A. (2012). Which "key to all mythologies" about the self? A note on where the illusions of transcendence come from and how to resist them. In S. Prosser & F. Recanati (Eds.), *Immunity to error through misidentification* (pp. 22–45). Cambridge: Cambridge University Press.

Colombo, M. (2013). Constitutive relevance and the personal/subpersonal distinction. *Philosophical Psychology*, *26*, 547–570.

Cosmides, L., & Tooby, J. (1992). Cognitive adaptations for social exchange. In J. Barkow, L. Cosmides, & J. Tooby (Eds.), *The adapted mind* (pp. 163–228). New York: Oxford University Press.

Crone, K. (2005). *Fichtes Theorie konkreter Subjektivität: Untersuchungen zur Wissenschaftslehre nova methodo, Neue Studien zur Philosophie* (Vol. 18). Göttingen: Vandenhoeck & Ruprecht.

Crowther, T. M. (2006). Two conceptions of conceptualism and nonconceptualism. *Erkenntnis*, *65*(2), 245–276.

Cussins, A. (2003). Content, conceptual content, and nonconceptual content. In Y. H. Gunther (Ed.), *Essays on nonconceptual content* (pp. 133–163). Cambridge, MA: MIT Press. (Original work published in 1990.)

Dally, J. M., Emery, N. J., & Clayton, N. S. (2006). Food-caching western scrub-jays keep track of who was watching when. *Science*, *312*(5780), 1662–1665.

Dapretto, M., Davies, M. S., Pfeifer, J. H., Scott, A. A., Sigman, M., Bookheimer, S. Y., (2005). Understanding emotions in others: Mirror neuron dysfunction in children with autism spectrum disorders. *Nature Neuroscience*, *9*(1), 28–30.

D'Argembeau, A., Collette, F., Van der Linden, M., Laureys, S., Del Fiore, G., Degueldre, C., (2005). Self-referential reflective activity and its relationship with rest: A PET study. *NeuroImage*, *25*, 616–624.

Davidson, D. (1975). Thought and talk. In S. Guttenplan (Ed.), *Mind and language: Wolfson College Lectures, 1974* (pp. 7–23). Oxford: Oxford University Press.

Dennett, D. C. (1987). *The intentional stance.* Cambridge, MA: MIT Press.

de Vignemont, F. (2007). Habeas corpus: The sense of ownership of one's own body. *Mind & Language*, *22*(4), 427–449.

de Vignemont, F. (2012). Bodily immunity to error. In F. Recanati & S. Prosser (Eds.), *Immunity to error through misidentification* (pp. 224–246). Cambridge: Cambridge University Press.

de Villiers, J., & Pyers, J. (1997). Complementing cognition: The relationship between language and theory of mind. In *Proceedings of the 21st Annual Boston University Conference on Language Development* (pp. 136–147). Somerville, MA: Cascadilla Press.

de Villiers, P. A. (2005). The role of language in theory-of-mind development: What deaf children tell us. In J. W. Astington & J. A. Baird (Eds.), *Why language matters for theory of mind* (pp. 266–297). Oxford: Oxford University Press.

Dienes, Z., & Perner, J. (1999). A theory of implicit and explicit knowledge. *Behavioral and Brain Sciences, 22*(5), 735–808.

Dokic, J. (2003). The sense of ownership: An analogy between sensation and action. In J. Roessler & N. Eilan (Eds.), *Agency and self-awareness* (pp. 321–344). Oxford: Oxford University Press.

Dow, J. M. (2012). On the joint engagement of persons: Self-consciousness, the symmetry thesis, and person perception. *Philosophical Psychology, 25*(1), 1–27.

Dretske, F. (1981). *Knowledge and the flow of information.* Cambridge, MA: MIT Press.

Dummett, M. (1978). *Truth and other enigmas.* Cambridge, MA: Harvard University Press.

Dummett, M. (1993). Thought and language. In M. Dummett (Ed.), *Origins of analytic philosophy* (pp. 127–161). Cambridge, MA: Harvard University Press.

Emery, N. J., & Clayton, N. S. (2001). Effects of experience and social context on prospective caching strategies by scrub jays. *Nature, 414*(6862), 443–446.

Evans, G. (1982). *The varieties of reference.* Oxford: Oxford University Press.

Feinman, S. (1982). Social referencing in infancy. *Merrill-Palmer Quarterly, 28*(4), 445–470.

Fernández, J. (2010). Thought insertion and self-knowledge. *Mind & Language, 25,* 66–88.

Ferrari, P. F., Visalberghi, E., Paukner, A., Fogassi, L., Ruggiero, A., & Suomi, S. J. (2006). Neonatal imitation in rhesus macaques. *PLoS Biology, 4*(9), 1501.

Fichte, J. G., Krause, K. C., & Fuchs, E. (1982). *Wissenschaftslehre nova methodo.* Hamburg: Felix Meiner. (Original work published in 1794–95.)

Fiebich, A., & Coltheart, M. (in press). Various ways to understand other minds: Towards a pluralistic approach to the explanation of social understanding. *Mind and Language.*

Field, J. (1976). Relation of young infants' reaching behavior to stimulus distance and solidity. *Developmental Psychology, 12,* 444–448.

Flanagan, O., Jr. (1992). *Consciousness reconsidered.* Cambridge, MA: MIT Press.

Flavell, J. H., Everett, B. A., Croft, K., & Flavell, E. R. (1981). Young children's knowledge about visual perception: Further evidence for the level 1–level 2 distinction. *Developmental Psychology, 17*(1), 99–103.

Flavell, J. H., Flavell, E. R., & Green, F. L. (1983). Development of the appearance—reality distinction. *Cognitive Psychology, 15*(1), 95–120.

Flombaum, J. I., & Santos, L. R. (2005). Rhesus monkeys attribute perceptions to others. *Current Biology, 15,* 447–452.

Fodor, J. A. (1983). *The modularity of mind.* Cambridge, MA: MIT Press.

Fodor, J. A. (1998). *Concepts: Where cognitive science went wrong.* Oxford: Oxford University Press.

Frank, M. (1991a). *Selbstbewusstseinstheorien von Fichte bis Sartre.* Berlin: Suhrkamp.

Frank, M. (1991b). *Selbstbewusstsein und Selbsterkenntnis: Essays zur analytischen Philosophie der Subjektivität.* Stuttgart: Philipp Reclam jun.

Frank, M. (2002). Self-consciousness and self-knowledge: On some difficulties with the reduction of subjectivity. *Constellations, 9*(3), 390–408.

Frank, M. (2007). Non-objectal subjectivity. *Journal of Consciousness Studies, 14,* 152–173.

Frith, C. D. (1992). *The cognitive psychology of schizophrenia.* Hillsdale, NJ: Erlbaum.

Frith, U. (2003). *Autism: Explaining the enigma.* Cambridge, MA: Blackwell.

Frith, U., & de Vignemont, F. (2005). Egocentrism, allocentrism, and Asperger syndrome. *Consciousness and Cognition, 14*(4), 719–738.

Frith, U., & Happé, F. (1999). Theory of mind and self-consciousness: What is it like to be autistic? *Mind & Language, 14*(1), 82–89.

Gallagher, S. (1986). Body image and body schema: A conceptual clarification. *Journal of Mind and Behavior, 7,* 541–554.

Gallagher, S. (2000). Philosophical conceptions of the self: Implications for cognitive science. *Trends in Cognitive Sciences, 4*(1), 14–21.

Gallagher, S. (2003). Bodily self-awareness and object-perception. *Theoria et Historia Scientiarum: International Journal for Interdisciplinary Studies, 7,* 53–68.

Gallagher, S. (2004). Neurocognitive models of schizophrenia: A neurophenomenological critique. *Psychopathology, 37*(1), 8–19.

Gallagher, S. (2007). Simulation trouble. *Social Neuroscience*, *2*(3), 353–365.

Gallagher, S., & Hutto, D. (2008). Understanding others through primary interaction and narrative practice. In J. Zlatev, T. Racine, C. Sinha, & E. Itkonen (Eds.), *The shared mind: Perspectives on intersubjectivity* (pp. 17–38). Amsterdam: John Benjamins.

Gallagher, S., & Zahavi, D. (2010). Phenomenological approaches to self-consciousness. In Edward N. Zalta (Ed.), *The Stanford encyclopedia of philosophy* (Winter 2010 Ed.), http://plato.stanford.edu/archives/win2010/entries/self-consciousness-phenom enological.

Gallese, V. (2001). The shared manifold hypothesis: From mirror neurons to empathy. *Journal of Consciousness Studies*, *8*(5–7), 33–50.

Gallese, V., & Goldman, A. (1998). Mirror neurons and the simulation theory of mind-reading. *Trends in Cognitive Sciences*, *2*(12), 493–501.

Gallese, V., Keysers, C., & Rizzolatti, G. (2004). A unifying view of the basis of social cognition. *Trends in Cognitive Sciences*, *8*(9), 396–403.

Gallup, G. G., Jr. (1979). Self-recognition in chimpanzees and man: A developmental and comparative perspective. In M. Lewis & L. Rosenblum (Eds.), *Genesis of behavior: The child and its family* (Vol. 4, pp. 107–126). New York: Plenum.

Georgieff, N., & Jeannerod, M. (1998). Beyond consciousness of external reality: A "who" system for consciousness of action and self-consciousness. *Consciousness and Cognition*, *7*(3), 465–477.

Gergely, G., Bekkering, H., & Király, I. (2002). Developmental psychology: Rational imitation in preverbal infants. *Nature*, *415*(6873), 755.

Gibson, J. J. (1979). *The ecological approach to perception*. Boston: Houghton Mifflin.

Ginsborg, H. (2011a). Primitive normativity and skepticism about rules. *Journal of Philosophy*, *108*(5), 227–254.

Ginsborg, H. (2011b). Perception, generality, and reasons. In A. Reisner & A. Steglich-Petersen (Eds.), *Reasons for belief* (pp. 131–157). Cambridge: Cambridge University Press.

Gloy, K. (1998). *Bewusstseinstheorien: Zur Problematik und Problemgeschichte des Bewusstseins und Selbstbewusstseins*. Freiburg: K. Alber.

Goldberg, I. I., Harel, M., & Malach, R. (2006). When the brain loses its self: Prefrontal inactivation during sensorimotor processing. *Neuron*, *50*(2), 329–339.

Goldman, A. I. (1992). In defense of the simulation theory. *Mind & Language*, *7*(1–2), 104–119.

Goldman, A. I. (2006). *Simulating minds: The philosophy, psychology, and neuroscience of mindreading*. Oxford: Oxford University Press.

Gopnik, A., & Meltzoff, A. N. (1997). *Words, thoughts, and theories*. Cambridge, MA: MIT Press.

Gopnik, A., & Wellman, H. M. (1992). Why the child's theory of mind really is a theory. *Mind & Language, 7*(1–2), 145–171.

Gordon, R. (1986). Folk psychology as simulation. *Mind & Language, 1*, 158–171.

Gordon, R. (2009). Folk psychology as mental simulation. In Edward N. Zalta (Ed.), *The Stanford encyclopedia of philosophy* (Fall 2009 Ed.), http://plato.stanford.edu/archives/fall2009/entries/folkpsych-simulation.

Grice, H. P. (1957). Meaning. *Philosophical Review, 66*(3), 377–388.

Griffiths, D., Dickinson, A., & Clayton, N. (1999). Episodic memory: What can animals remember about their past? *Trends in Cognitive Sciences, 3*(2), 74–80.

Habermas, J. (1999). From Kant to Hegel and back again: The move towards detranscendentalization. *European Journal of Philosophy, 7*, 129–157.

Hamilton, A. (2013). *The self in question: Memory, the body, and self-consciousness*. Basingstoke, Hampshire: Palgrave Macmillan.

Hampton, R. (2001). Rhesus monkeys know when they remember. *Proceedings of the National Academy of Sciences of the United States of America, 98*, 5359–5362.

Hampton, R. (2005). Can rhesus monkeys discriminate between remembering and forgetting? In H. Terrace & J. Metcalfe (Eds.), *The missing link in cognition: Origins of self-reflective consciousness* (pp. 272–295). Oxford: Oxford University Press.

Hanna, R. (2008). Kantian non-conceptualism. *Philosophical Studies, 137*(1), 41–64.

Happé, F. (2003). Theory of mind and the self. *Annals of the New York Academy of Sciences, 1001*, 134–144.

Heal, J. (1986). Replication and functionalism. In J. Butterfield (Ed.), *Language, mind, and logic* (pp. 135–150). Cambridge: Cambridge University Press.

Heck, R. G., Jr. (2000). Nonconceptual content and the "space of reasons." *Philosophical Review, 109*(4), 483–523.

Heck, R. G., Jr. (2007). Are there different kinds of content? In B. McLaughlin & J. Cohen (Eds.), *Contemporary debates in philosophy of mind* (pp. 117–138). Malden, MA: Blackwell.

Heidegger, M. (1989). *Die Grundprobleme der Phänomenologie, Gesamtausgabe* (Vol. 24). Frankfurt am Main: Vittorio Klostermann.

Held, R. (1961). Exposure-history as a factor in maintaining stability of perception and coordination. *Journal of Nervous and Mental Disease, 132*(1), 26–32.

Henrich, D. (1967). *Fichtes ursprüngliche Einsicht*. Frankfurt am Main: Klostermann.

Henry, M. (1975). *Philosophy and phenomenology of the body*. Heidelberg: Springer.

Heyes, C. (1994). Reflections on self-recognition in primates. *Animal Behaviour, 47*(4), 909–919.

Heyes, C. (2009). Where do mirror neurons come from? *Neuroscience and Biobehavioral Reviews, 34*(4), 575–583.

Heyes, C., & Dickinson, A. (1990). The intentionality of animal action. *Mind & Language, 5*(1), 87–103.

Hobson, P. (2002). *The cradle of thought: Explorations of the origins of thinking*. Oxford: Macmillan.

Hopp, W. (2011). *Perception and knowledge: A phenomenological account*. Cambridge: Cambridge University Press.

Hornsby, J. (2000). Personal and subpersonal: A defence of Dennett's early distinction. *Philosophical Explorations, 3*, 6–24.

Hume, D. (1967). *A treatise of human nature*. London: Oxford University Press. (Original work published in 1888.)

Hurlburt, R. T., & Schwitzgebel, E. (2007). *Describing inner experience? Proponent meets skeptic*. Cambridge, MA: MIT Press.

Hurley, S. L. (1997). Nonconceptual self-consciousness and agency: Perspective and access. *Communication and Cognition (Part 1 of Special Issue: Approaching Consciousness), 30*(3–4), 207–248.

Hurley, S. L. (1998). *Consciousness in action*. Cambridge, MA: Harvard University Press.

Hurley, S. L. (2006). Making sense of animals. In S. Hurley & M. Nudds (Eds.), *Rational animals* (pp. 231–256). Oxford: Oxford University Press.

Hurley, S. L. (2008). The shared circuits model (SCM): How control, mirroring, and simulation can enable imitation, deliberation, and mindreading. *Behavioral and Brain Sciences, 31*(01), 1–22.

Hutto, D. (2008). The narrative practice hypothesis: Clarifications and implications. *Philosophical Explorations, 11*(3), 175–192.

Iacoboni, M., Woods, R. P., Brass, M., Bekkering, H., Mazziotta, J. C., & Rizzolatti, G. (1999). Cortical mechanisms of human imitation. *Science, 286*(5449), 2526.

Jeannerod, M., & Pacherie, E. (2004). Agency, simulation, and self-identification. *Mind & Language, 19*(2), 113–146.

Johnson, M. H., & Morton, J. (1991). *Biology and cognitive development: The case of face recognition.* Oxford: Blackwell.

Kaminski, J., Call, J., & Tomasello, M. (2008). Chimpanzees know what others know, but not what they believe. *Cognition, 109*(2), 224–234.

Kant, I. (2009). *Critique of pure reason* (15th ed.) (P. Guyer & A. W. Wood, Trans.). Cambridge: Cambridge University Press. (Original work published in 1781 [A edition] and 1787 [B edition].)

Kapitan, T. (2006). Indexicality and self-awareness. In U. Kriegel & K. Williford (Eds.), *Self-representational approaches to consciousness* (pp. 379–408). Cambridge, MA: MIT Press.

Kaplan, D. (1977). Demonstratives. In J. Almog, J. Perry, & H. Wettstein (Eds.), *Themes from Kaplan* (pp. 481–563). Oxford: Oxford University Press.

Karmiloff-Smith, A. (1984). Children's problem solving. *Advances in Developmental Psychology, 3*, 39–90.

Karmiloff-Smith, A. (1996). *Beyond modularity* (1st paperback ed.). Cambridge, MA: MIT Press.

Kazak, S., Collis, G. M., & Lewis, V. (1997). Can young people with autism refer to knowledge states? Evidence from their understanding of "know" and "guess." *Journal of Child Psychology and Psychiatry, and Allied Disciplines, 38*(8), 1001–1009.

Kellman, P. J., & Spelke, E. S. (1983). Perception of partly occluded objects in infancy. *Cognitive Psychology, 15*(4), 483–524.

Kelly, S. D. (2001). The non-conceptual content of perceptual experience: Situation dependence and fineness of grain. *Philosophy and Phenomenological Research, 62*(3), 601–608.

Keysers, C., & Gazzola, V. (2009). Expanding the mirror: Vicarious activity for actions, emotions, and sensations. *Current Opinion in Neurobiology, 19*, 666–671.

Kirsh, D. (1990). When is information explicitly represented. In P. Hanson (Ed.), *Information, content, and meaning* (pp. 340–365). Vancouver: UBC Press.

Kirsh, D. (2009). Knowledge, explicit vs. implicit. In T. Bayne, A. Cleeremans, & P. Wilken (Eds.), *Oxford companion to consciousness* (pp. 397–402). Oxford: Oxford University Press.

Klinnert, M., Campos, J. J., Sorce, J. F., Emde, R. N., & Svejda, M. (1983). Emotions as behavior regulators: Social referencing in infancy. *Emotions in Early Development, 2*, 57–86.

Koffka, K. (1935). *Principles of gestalt psychology*. London: Lund Humphries.

Kohler, I. (1964). The formation and transformation of the perceptual world. *Psychological Issues, 3*(12), 1–173.

Kohler, W. (1947). *Gestalt psychology*. New York: Liveright. (Original work published in 1929.)

Kornell, N. (2014). Where is the "meta" in animal metacognition? *Journal of Comparative Psychology, 128*(2), 143–149.

Kovács, Á. M., Téglás, E., & Endress, A. D. (2010). The social sense: Susceptibility to others' beliefs in human infants and adults. *Science, 330*(6012), 1830–1834.

Kriegel, U. (2009). *Subjective consciousness: A self-representational theory*. Oxford: Oxford University Press.

Lailach-Hennrich, A. (2011). *Ich und die Anderen: Quellen und Studien zur Philosophie*. Berlin: De Gruyter.

Legerstee, M. (1992). A review of the animate—inanimate distinction in infancy: Implications for models of social and cognitive knowing. *Early Development & Parenting, 1*(2), 59–67.

Legrand, D. (2006). The bodily self: The sensori-motor roots of pre-reflective self-consciousness. *Phenomenology and the Cognitive Sciences, 5*(1), 89–118.

Legrand, D. (2007). Pre-reflective self-as-subject from experiential and empirical perspectives. *Consciousness and Cognition, 16*(3), 583–599.

Leslie, A. M., German, T. P., & Polizzi, P. (2005). Belief—desire reasoning as a process of selection. *Cognitive Psychology, 50*, 45–85.

Leslie, A. M., Hirschfeld, L. A., & Gelman, S. A. (1994). *Mapping the mind: Domain specificity in cognition and culture*. Cambridge: Cambridge University Press.

Leslie, A. M., & Polizzi, P. (1998). Inhibitory processing in the false belief task: Two conjectures. *Developmental Science, 1*, 247–253.

Lewis, D. (1979). Attitudes *de dicto* and *de se*. *Philosophical Review, 88*, 513–543.

Lewis, M., Sullivan, M. W., Stanger, C., & Weiss, M. (1989). Self development and self-conscious emotions. *Child Development, 60*(1), 146–156.

Lohmann, H., & Tomasello, M. (2003). The role of language in the development of false belief understanding: A training study. *Child Development, 74*(4), 1130–1144.

Lurz, R. (2011). *Mindreading animals: The debate over what animals know about other minds*. Cambridge, MA: MIT Press.

Mack, A., & Rock, I. (1998). *Inattentional blindness*. Cambridge, MA: MIT Press.

Mandler, J. (2004). *The foundations of mind: Origins of conceptual thought.* Oxford: Oxford University Press.

Marten, K., & Psarakos, S. (1994). Evidence of self-awareness in the bottlenose dolphin (Tursiops truncate). In S. Parker, M. Boccia, & R. Mitchell (Eds.), *Self-awareness in animals and humans: Developmental perspectives* (pp. 361–379). Cambridge: Cambridge University Press.

Martin, M. G. F. (1995). Bodily awareness: A sense of ownership. In J. L. Bermúdez, N. Eilan, & A. Marcel (Eds.), *The body and the self* (pp. 267–289). Cambridge, MA: MIT Press.

McDowell, J. (1994). *Mind and world.* Cambridge, MA: Harvard University Press.

McDowell, J. (1998). Reductionism and the first person. In McDowell, *Mind, value, and reality* (pp. 359–382). Cambridge, MA: Harvard University Press.

Mead, G. H. (1934). *Mind, self, and society.* Chicago: University of Chicago Press.

Meeks, R. (2006). Why nonconceptual content cannot be immune to error through misidentification. *European Review of Philosophy, 6,* 83–102.

Mehler, J., Jusczyk, P., & Lambertz Nilofar, G. (1988). A precursor of language acquisition in young infants. *Cognition, 29*(2), 143–178.

Meltzoff, A. N. (1990). Towards a developmental cognitive science: The implications of cross-modal matching and imitation for the development of representation and memory in infancy. *Annals of the New York Academy of Sciences, 608,* 1–37.

Meltzoff, A. N. (1995). Understanding the intentions of others: Re-enactment of intended acts by 18-month-old children. *Developmental Psychology, 31*(5), 838–850.

Meltzoff, A. N., & Moore, K. (1977). Imitation of facial and manual gestures by newborn infants. *Science, 198,* 75–78.

Merleau-Ponty, M. (1962). *Phenomenology of perception* (C. Smith, Trans.). London: Routledge. (Original work published in 1945.)

Millikan, R. G. (1990). The myth of the essential indexical. *Noûs, 24*(5), 723–734.

Millikan, R. G. (1993). Content and vehicle. In N. Eilan, R. McCarthy, & B. Brewer (Eds.), *Spatial representation* (pp. 256–268). Oxford: Blackwell.

Milner, A. D., & Goodale, M. A. (1995). *The visual brain in action.* Oxford: Oxford University Press.

Mitchell, R. W. (1993). Mental models of mirror-self-recognition: Two theories. *New Ideas in Psychology, 11,* 295–325.

Mitchell, R. W. (2002). Kinesthetic-visual matching, imitation, and self-recognition. In M. Bekoff, C. Allen, & G. M. Burhgardt (Eds.), *The cognitive animal: Empirical and theoretical perspectives on animal cognition* (pp. 345–351). Cambridge, MA: MIT Press.

Moll, H., & Tomasello, M. (2006). Level 1 perspective-taking at 24 months of age. *British Journal of Developmental Psychology, 24*(3), 603–613.

Moran, R. (2001). *Authority and estrangement: An essay on self-knowledge.* Princeton: Princeton University Press.

Morin, A. (2005). Possible links between self-awareness and inner speech: Theoretical background, underlying mechanisms, and empirical evidence. *Journal of Consciousness Studies, 12*(4–5), 115–134.

Musholt, K. (2012). Self-consciousness and intersubjectivity. *Grazer Philosophische Studien, 84*, 63–89.

Musholt, K. (2013a). Self-consciousness and nonconceptual content. *Philosophical Studies, 163*(3), 649–672.

Musholt, K. (2013b). A philosophical perspective on the relation between cortical midline structures and the self. *Frontiers in Human Neuroscience, 7*, 1–11.:

Nelson, K. (2007). *Young minds in social worlds: Experience, meaning, and memory.* Cambridge, MA: Harvard University Press.

Newen, A., & Schlicht, T. (2009). Understanding other minds: A criticism of Goldman's simulation theory and an outline of the person model theory. *Grazer Philosophische Studien, 79*(1), 209–242.

Noë, A. (2002). Is perspectival self-consciousness non-conceptual? *Philosophical Quarterly, 52*(207), 185–194.

Noë, A. (2005). Anti-intellectualism. *Analysis, 65*(4), 278–289.

Northoff, G. (2011). Self and brain: What is self-related processing? *Trends in Cognitive Sciences, 15*(5), 186–187.

Northoff, G. (2013). *Unlocking the brain.* New York: Oxford University Press.

Northoff, G., & Bermpohl, F. (2004). Cortical midline structures and the self. *Trends in Cognitive Sciences, 8*, 102–107.

O'Brien, L. (2007). *Self-knowing agents.* Oxford: Oxford University Press.

O'Brien, L. (2012). Action and immunity to error through misidentification. In S. Prosser & F. Recanati (Eds.), *Immunity to error through misidentification* (pp. 124–143). Cambridge: Cambridge University Press.

Onishi, K. H., & Baillargeon, R. (2005). Do 15-month-old infants understand false beliefs? *Science, 308*, 255–258.

Opfer, J. E., & Gelman, S. A. (2010). Development of the animate—inanimate distinction. In U. Goswami (Ed.), *Wiley-Blackwell handbook of childhood cognitive development* (pp. 213–238). Oxford: Blackwell.

O'Regan, J. K., & Noë, A. (2002). A sensorimotor account of vision and visual consciousness. *Behavioral and Brain Sciences, 24*(5), 939–973.

O'Shaughnessy, B. (1995). Proprioception and the body image. In N. Eilan & J.-L. Bermúdez (Eds.), *The body and the self* (pp. 175–203). Cambridge, MA: MIT Press.

Pauen, M. (2000). Selbstbewußtsein: Ein metaphysisches Relikt? In A. Newen & K. Vogeley (Eds.), *Selbst und Gehirn: Menschliches Selbstbewußtsein und seine neurobiologischen Grundlagen* (pp. 101–121). Paderborn: Mentis.

Pauen, M. (2012). The second-person perspective. *Inquiry, 55*(1), 33–49.

Peacocke, C. (1992). *A study of concepts.* Cambridge, MA: MIT Press.

Peacocke, C. (1999). *Being known.* Oxford: Oxford University Press.

Peacocke, C. (2001). Does perception have a nonconceptual content? *Journal of Philosophy, 98*(5), 239–264.

Peacocke, C. (2012). Explaining de se phenomena. In S. Prosser & F. Recanati (Eds.), *Immunity to error through misidentification* (pp. 144–157). Cambridge: Cambridge University Press.

Perner, J. (1991). *Understanding the representational mind.* Cambridge, MA: MIT Press.

Perner, J., Frith, U., Leslie, A. M., & Leekam, S. R. (1989). Exploration of the autistic child's theory of mind: Knowledge, belief, and communication. *Child Development, 60*(3), 689–700.

Perner, J., Leekam, S. R., & Wimmer, H. (1987). Three-year-olds' difficulty with false belief: The case for a conceptual deficit. *British Journal of Developmental Psychology, 5*(2), 125–137.

Perner, J., & Ruffman, T. (2005). Infant's insight into the mind: How deep? *Science, 308*, 214–216.

Perry, J. (1979). The essential indexical. *Noûs, 13*, 3–21.

Perry, J. (1986). Thought without representation. *Proceedings of the Aristotelian Society, Supplementary Volumes, 60*, 137–166.

Perry, J. (1998a). Indexicals, contexts, and unarticulated constituents. In *Proceedings of the 1995 CSLI-Amsterdam Logic, Language and Computation Conference* (pp. 1–16). Stanford: CSLI Publications.

Perry, J. (1998b). Myself and I. In M. Stamm (Ed.), *Philosophie in synthetischer Absicht (Festschrift für Dieter Henrich)* (pp. 83–103). Stuttgart: Klett-Cotta.

Perry, J. (2000). *The problem of the essential indexical and other essays* (expanded ed.). Stanford: CSLI Publications.

Perry, J. (2002). *Idenity, personal identity, and the self*. Indianapolis: Hackett.

Peterson, C. C., & Siegal, M. (1995). Deafness, conversation and theory of mind. *Journal of Child Psychology and Psychiatry, and Allied Disciplines, 36*(3), 459–474.

Pettit, P. (2003). Looks as powers. *Philosophical Issues, 13*(1), 221–252.

Phillips, A. T., Wellman, H. M., & Spelke, E. S. (2002). Infants' ability to connect gaze and emotional expression to intentional action. *Cognition, 85*(1), 53–78.

Phillips, W., Baron-Cohen, S., & Rutter, M. (1998). Understanding intention in normal development and in autism. *British Journal of Developmental Psychology, 16*(3), 337–348.

Plotnik, J. M., De Waal, F. B., & Reiss, D. (2006). Self-recognition in an Asian elephant. *Proceedings of the National Academy of Sciences, 103*(45), 17053–17057.

Poellner, P. (2003). Non-conceptual content, experience and the self. *Journal of Consciousness Studies, 10*(2), 32–57.

Pothast, U. (1971). *Über einige Fragen der Selbstbeziehung*. Frankfurt am Main: Klostermann.

Preston, S. D., & de Waal, F. B. (2003). Empathy: Its ultimate and proximate bases. *Behavioral and Brain Sciences, 25*(1), 1–20.

Prior, H., Schwarz, A., & Güntürkün, O. (2008). Mirror-induced behavior in the magpie (Pica pica): Evidence of self-recognition. *PLoS Biology, 6*(8), e202.

Proust, J. (2006). Rationality and meta-cognition in non-human animals. In S. Hurley & M. Nudds (Eds.), *Rational animals?* (pp. 247–274). Oxford: Oxford University Press.

Qin, P., & Northoff, G. (2011). How is our self related to midline regions and the default-mode network? *NeuroImage, 57*, 1221–1233.

Raichle, M. E., MacLeod, A. M., Snyder, A. Z., Powers, W. J., Gusnard, D. A., & Shulman, G. L. (2001). A default mode of brain function. *Proceedings of the National Academy of Sciences of the United States of America, 98*, 676–682.

Rakoczy, H. (2008). Du, Ich, Wir: Zur Entwicklung sozialer Kognition bei Mensch und Tier. In R. Schubotz (Ed.), *Other minds*. Paderborn: Mentis.

Recanati, F. (2007). *Perspectival thought: A plea for (moderate) relativism*. Oxford: Oxford University Press.

Recanati, F. (2009). De re and de se. *Dialectica, 63*(3), 249–269.

Recanati, F. (2012). Immunity to error through misidentification: What it is and where it comes from. In S. Prosser & F. Recanati (Eds.), *Immunity to error through misidentification* (pp. 180–201). Cambridge: Cambridge University Press.

Repacholi, B., & Gopnik, A. (1997). Early understanding of desires: Evidence from 14- and 18-month-olds. *Developmental Psychology, 33*(1), 12–21.

Rizzolatti, G., Fadiga, L., Gallese, V., & Fogassi, L. (1996). Premotor cortex and the recognition of motor actions. *Cognitive Brain Research, 3*(2), 131–141.

Robinson, E., & Apperly, I. A. (2001). Children's difficulties with partial representations in ambiguous messages and referentially opaque contexts. *Cognitive Development, 16*, 595–615.

Rochat, P., & Hespos, S. J. (1997). Differential rooting response by neonates: Evidence for an early sense of self. *Early Development & Parenting, 6*(34), 105–112.

Rochat, P. (2009). *Others in mind: Social origins of self-consciousness*. Cambridge: Cambridge University Press.

Rödl, S. (2007). *Self-consciousness*. Cambridge, MA: Harvard University Press.

Rosefeldt, T. (2000). Sich setzen, oder Was ist eigentlich das Besondere an Selbstbewusstsein? John Perry hilft, eine Debatte zwischen Henrich und Tugendhat zu klären. *Zeitschrift für Philosophische Forschung, 54*(3), 425–444.

Rosefeldt, T. (2004). Is knowing-how simply a case of knowing-that? *Philosophical Investigations, 27*(4), 370–379.

Rosenthal, D. M. (1986). Two concepts of consciousness. *Philosophical Studies, 49*(3), 329–359.

Roskies, A. L. (2008). A new argument for nonconceptual content. *Philosophy and Phenomenological Research, 76*(3), 633–659.

Ryle, G. (1949). *The concept of mind*. London: Hutchinson.

Ryle, G. (1945). Knowing how and knowing that: The presidential address. *Proceedings of the Aristotelian Society, 56*, 1–16.

Santos, L., Nissen, A., & Ferrugia, J. (2006). Rhesus monkeys, Macaca mulatta, know what others can and cannot hear. *Animal Behaviour, 71*, 1175–1181.

Sartre, J. P. (1966). *Being and nothingness* (H. E. Barnes, Trans.). New York: Washington Square Press. (Original work published in 1943.)

Schacter, D. L. (1987). Implicit memory: History and current status. *Journal of Experimental Psychology: Learning, Memory, and Cognition, 13*(3), 501–518.

Schear, J. K. (2009). Experience and self-consciousness. *Philosophical Studies, 144*(1), 95–105.

Schmidt, M. F. H., Rakoczy, H., & Tomasello, M. (2011). Young children attribute normativity to novel actions without pedagogy or normative language. *Developmental Science, 14*(3), 530–539.

Schmidt, M., & Tomasello, M. (2012). Young children enforce social norms. *Current Directions in Psychological Science, 21*, 232–236.

Schneider, F., Bermpohl, F., Heinzel, A., Rotte, M., Walter, M., Tempelmann, C., (2008). The resting brain and our self: Self-relatedness modulates resting state neural activity in cortical midline structures. *Neuroscience, 157*, 120–131.

Schwenkler, J. (2014). Vision, self-location, and the phenomenology of the "point of view." *Noûs, 48*(1), 137–155.

Shields, W., Smith, J., & Washburn, D. (1997). Uncertain responses by humans and rhesus monkeys (Macaca mulatta) in a psychophysical same—different task. *Journal of Experimental Psychology: General, 126*, 147–164.

Shoemaker, S. (1968). Self-reference and self-awareness. *Journal of Philosophy, 65*(19), 555–567.

Shoemaker, S. (1996). *The first-person perspective and other essays*. Cambridge: Cambridge University Press.

Shoemaker, S. (2003). *Identity, cause, and mind: Philosophical essays*. Oxford: Oxford University Press.

Simons, D. J., & Chabris, C. F. (1999). Gorillas in our midst: Sustained inattentional blindness for dynamic events. *Perception, 28*, 1059–1074.

Simons, D. J., & Rensink, R. A. (2005). Change blindness: Past, present, and future. *Trends in Cognitive Sciences, 9*(1), 16–20.

Smith, J. (2005). Studies of uncertainty monitoring and meta-cognition in animals and humans. In H. Terrace & J. Metcalfe (Eds.), *The missing link in cognition: Origins of self-reflective consciousness* (pp. 242–271). Oxford: Oxford University Press.

Smith, J., Schull, J., Strote, J., McGee, K., Egnor, R., & Erb, L. (1995). The uncertain response in the bottlenosed dolphin (Tursiops truncatus). *Journal of Experimental Psychology: General, 124*, 391–408.

Sodian, B. (1991). The development of deception in young children. *British Journal of Developmental Psychology, 9*(1), 173–188.

Sodian, B. (1994). Early deception and the conceptual continuity claim. In C. Lewis & P. Mitchell (Eds.), *Children's early understanding of mind: Origins and development* (pp. 385–401). Hove: Erlbaum.

Son, L., & Kornell, N. (2005). Meta-confidence judgments in rhesus macaques: Explicit versus implicit mechanisms. In H. Terrace & J. Metcalfe (Eds.), *The missing link in cognition: Origins of self-reflective consciousness* (pp. 296–320). Oxford: Oxford University Press.

Spaulding, S. (2011). A critique of embodied simulation. *Review of Philosophy and Psychology*, *2*(3), 579–599.

Spelke, E. S. (1988). Where perceiving ends and thinking begins: The apprehension of objects in infancy. In A. Yonas (Ed.), *Minnesota Symposium on Child Psychology: Perception* (Vol. 20, p. 321). Hillsdale, NJ: Erlbaum.

Spelke, E. S. (1990). Principles of object perception. *Cognitive Science*, *14*(1), 29–56.

Spelke, E. S. (2009). Comments. In M. Tomasello (Ed.), *Why we cooperate* (pp. 149–172). Cambridge, MA: MIT Press.

Stalnaker, R. (1998). What might nonconceptual content be? *Philosophical Issues*, *9*, 339–352.

Stanley, J., & Williamson, T. (2001). Knowing how, knowing that. *Journal of Philosophy*, *98*(8), 411–444.

Stephens, G. L., & Graham, G. (2000). *When self-consciousness breaks: Alien voices and inserted thoughts*. Cambridge, MA: MIT Press.

Strawson, P. F. (1959). *Individuals: An essay in descriptive metaphysics*. London: Routledge.

Suarez, S. D., & Gallup, G. G. (1981). Self-recognition in chimpanzee and orangutans, but not gorillas. *Journal of Human Evolution*, *10*, 175–188.

Suddendorf, T., & Butler, D. (2013). The nature of visual self-recognition. *Trends in Cognitive Sciences*, *17*(3), 121–127.

Sun, R., Merrill, E., & Petersen, T. (2001). From implicit skills to explicit knowledge: A bottom-up model of skill learning. *Cognitive Science*, *25*(2), 203–244.

Surian, L., Caldi, S., & Sperber, D. (2007). Attribution of beliefs by 13-month-old infants. *Psychological Science*, *18*(7), 580–586.

Synofzik, M., Vosgerau, G., & Newen, A. (2008). Beyond the comparator model: A multifactorial two-step account of agency. *Consciousness and Cognition*, *17*(1), 219–239.

Thiel, U. (2011). *The early modern subject: Self-consciousness and personal identity from Descartes to Hume*. Oxford: Oxford University Press.

Tomasello, M. (1999). *The cultural origins of human cognition*. Cambridge, MA: Harvard University Press.

Tomasello, M. (2008). *Origins of human communication*. Cambridge, MA: MIT Press.

Tomasello, M. (2009). *Why we cooperate*. Cambridge, MA: MIT Press.

Tomasello, M., & Call, J. (1997). *Primate cognition.* New York: Oxford University Press.

Tomasello, M., Carpenter, M., Call, J., Behne, T., & Moll, H. (2005). Understanding and sharing intentions: The origins of cultural cognition. *Behavioral and Brain Sciences, 28*(5), 675–691.

Toribio, J. (2008). State versus content: The unfair trial of perceptual nonconceptualism. *Erkenntnis, 69*(3), 351–361.

Trevarthen, C. (1979). Communication and cooperation in early infancy: A description of primary intersubjectivity. In M. Bullowa (Ed.), *Before speech: The beginning of interpersonal communication* (pp. 321–347). Cambridge: Cambridge University Press.

Tschudin, A. (2006). Belief attribution tasks with dolphins: What social minds can reveal about animal rationality. In S. Hurley & M. Nudds (Eds.), *Rational animals* (pp. 413–436). Oxford: Oxford University Press.

Tugendhat, E. (1979). *Selbstbewusstsein und Selbstbestimmung.* Berlin: Suhrkamp.

Tuomela, R., & Miller, K. (1988). We-intentions. *Philosophical Studies, 53*(3), 367–389.

Tye, M. (2000). *Color, content, and consciousness.* Cambridge, MA: MIT Press.

Tye, M. (2006). Nonconceptual content, richness, and fineness of grain. In T. G. Szabo & J. Hawthorne (Eds.), *Perceptual experience* (pp. 504–530). Oxford: Oxford University Press.

Van Gulick, R. (2006). Mirror-mirror, is that all? In U. Kriegl & K. Williford (Eds.), *Self-representational approaches to consciousness* (pp. 11–40). Cambridge, MA: MIT Press.

von Holst, E., & Mittelstaedt, H. (1950). Das Reafferenzprinzip. *Naturwissenschaften, 37,* 464–476.

Vosgerau, G. (2009). *Mental representation and self-consciousness: From basic self-representation to self-related cognition.* Paderborn: Mentis.

Ward, D., Roberts, T., & Clark, A. (2011). Knowing what we can do: Actions, intentions, and the construction of phenomenal experience. *Synthese, 181*(3), 357–394.

Welch, R., & Warren, D. (1986). Intersensory interactions. In K. R. Boff & L. Kaufman (Eds.), *Handbook of perception and human performance* (pp. 25–36). London: Wiley-Blackwell.

Wellman, H. M., Cross, D., & Watson, J. (2001). Meta-analysis of theory-of-mind development: The truth about false belief. *Child Development, 72*(3), 655–684.

Wertheimer, M. (1958). Principles of perceptual organization. In D. Beardslee and M. Wertheimer (Eds.), *Readings in perception* (pp. 115–135). Princeton, NJ: Van Nostrand. (Original work published in 1923.)

Williams, D. M., & Happé, F. (2009). What did I say? versus what did I think? Attributing false beliefs to self amongst children with and without autism. *Journal of Autism and Developmental Disorders, 39*(6), 865–873.

Wimmer, H., & Perner, J. (1983). Beliefs about beliefs: Representation and constraining function of wrong beliefs in young children's understanding of deception. *Cognition, 13*(1), 103–128.

Wittgenstein, L. (1967). *Philosophical investigations* (3rd ed.) (G. E. M. Anscombe, Trans.). Oxford: Blackwell. (Original work published in 1953.)

Wittgenstein, L. (1958). *The blue and brown books.* Oxford: Blackwell.

Wood, A. W. (2006). Fichte's intersubjective I. *Inquiry, 49*(1), 62–79.

Wright, C. (1998). Self-knowledge: The Wittgensteinian legacy. In C. Wright, B. C. Smith, & C. Macdonald (Eds.), *Knowing our own minds* (pp. 13–45). Oxford: Clarendon Press.

Zahavi, D. (2005). *Subjectivity and selfhood: Investigating the first-person perspective.* Cambridge, MA: MIT Press.

Zahavi, D., & Kriegel, U. (forthcoming). For-me-ness: What it is and what it is not. In D. O. Dahlstrom, A. Elpidorou, & W. Hopp (Eds.), *Philosophy of mind and phenomenology: Conceptual and empirical approaches.* London: Routledge.

Zawidzki, T. (2013). *Mindshaping.* Cambridge, MA: MIT Press.

Zlatev, J. (2008). The co-evolution of intersubjectivity and bodily mimesis. In J. Zlatev, T. P. Racine, C. Sinha, & E. Itkonen (Eds.), *The shared mind: Perspectives on intersubjectivity* (pp. 215–244). Amsterdam: John Benjamins.

# Index